GEOGRAPHERS
Biobibliographical Studies

VOLUME 2

Edited by

T. W. Freeman and Philippe Pinchemel

on behalf of the

Commission on the History of Geographical Thought

of the International Geographical Union

and the International Union of the History

and Philosophy of Science

MANSELL 1978

© 1978   International Geographical Union

Mansell Information/Publishing Limited,
3 Bloomsbury Place, London WC1A 2QA
First published 1978

This volume forms part of the series *Studies in the
History of Geography* planned by the International
Geographical Union, Commission on the History of
Geographical Thought.   *Chairman*, Professor Philippe
Pinchemel, Institut de Géographie, Université de Paris,
191 rue Saint Jacques, 75005, Paris.   *Secretary*,
Professor T.W. Freeman, Department of Geography,
University of Manchester, Manchester M13 9PL.
Ordinary members: Professor Vladimir Annenkov, Geo-
graphical Institute, Academy of Sciences, Staromonetny
per 29, Moscow V-17; Professor Jozef Babicz, Institut
d'Histoire des Sciences et des Techniques, Polska
Akademia Nauk, Nowy Swiat 72, Warsaw; Professor George
Kish, Department of Geography, University of Michigan,
Ann Arbor, Michigan 48104; Professor Josef Schmithüsen,
Universität des Saarlandes, Geographisches Institut,
Universität Bau 11 IV, D-66, Saarbrücken 15, West
Germany; Professor Ichiro Suitsu, Institute of Geo-
graphy, Kyoto University, Yoshida, Sakyo-Ky-Kyoto,
Japan.   *Honorary members:* Mr Gerald Crone,
34 Cleveland Road, Ealing, London W13; Professor
Robert E. Dickinson, 636 West Roller Coaster Road,
Tucson, Arizona 85704; Professor R. Hartshorne,
Department of Geography, University of Wisconsin,
Madison, Wisconsin 53706; Professor Vintila Mihailescu,
Str. Muntii Tatral 2, Bucaresti 8.

International Standard Book Number 0   7201 0710 5

International Standard Serial Number 0308-6992

Enquiries concerning publications listed in the
'Bibliography and Sources' sections of the biobibliographies should be sent to the relevant publisher or
journal, and *not* to Mansell.

British Library Cataloguing in Publication Data
Geographers: biobibliographical studies.
(Studies in the history of geography.)
  Vol. 2:
  1. Geographers
  I. Freeman, Thomas Walter II. Pinchemel, Philippe
  III. Commission on the History of Geographical Thought
  IV. Series
  910'.92'2   G67

  ISBN 0-7201-0710-5

Printed by photolithography and bound at The Scolar
Press Limited, Ilkley, Yorkshire.

Contents

Introduction

Place and time are always of concern to geographers and their work is carried out in an environment constantly changing because of political, economic and social factors. In this second volume of *Geographers* there are studies of people who worked in places remote from Europe and North America, including Africa, Australia, India and New Zealand. The geographers range from Eratosthenes in ancient Egypt to Carl Sauer in modern California. Eratosthenes realized that there might be other worlds that he could never hope to see while Carl Sauer knew that in western America there had been other human environments preceding those of his own time, which had never been completely obliterated. The lives of some geographers are closely affected by the thought of their time: of those considered here, Keckermann shared the concern with the relationship between science and theology that was a live issue of the Reformation period, and Bose, a close associate of Mahatma Gandhi, was deeply involved in the political ferment of the Indian sub-continent.

The life and experiences of the individual geographers vary considerably: there may be a surprising story such as that of George Davidson, who, at an age when most people have long accepted retirement, found himself placed in a new university post in California which was to open opportunities for others, notably Carl Sauer who eventually became famous, though he had had to endure many difficulties in his earlier years. The ability to recognize and take advantage of an opportunity is evident in some lives. Sir Charles Cotton clearly saw what he could achieve for geomorphology and for his native New Zealand. Not the least of his merits was that he cared more for the work than for any honours or rewards that it might bring. There are elements of tragedy in some stories, such as that of Geoffrey Milne who had so short a time to work in East Africa on strictly scientific problems of immense human significance, or again in that of the Romanian geographer Vâlsan, whose career was marred by ill health but whose work, however uncompleted, was a fine heritage.

Universities and geographical societies provide support for the work of geographers but there have been many people who have acquired renown outside their confines. Because of political troubles in his homeland, the life of Strzelecki, the Pole, a man of aristocratic origin who spent many years pioneering in Australia, seems almost exotic, and is in marked contrast to that of Mary Somerville, happy among the intelligentsia of Edinburgh and then of London, though even she was eventually exiled to Italy for thirty years because of the precarious health and modest means of her husband. In his later years Strzelecki was able to give fine service to the Royal Geographical Society in London but Mary Somerville, though honoured by the Society in her late eighties, was debarred as a woman from even joining it.

Patrick Geddes held a university appointment at Dundee, which fortunately required his presence there for only part of the year, even though his initial career as a student was probably a record for brevity as it lasted less than one week nevertheless his eventual influence was wide, not only on geographers but also on planners and social scientists, for so many of the more enlightened ideas of an age of advance in scientific humanism were combined in him, that they overflowed in a ceaseless stream of talk and pamphlets.   A Scot whose work extended far beyond his homeland, his ideas were greater than his actual achievements.   His influence was marked, above all, on his son Arthur who gave inspiration to others, not least by his own fieldwork in Scotland and India.

As in other disciplines, geography may possess an 'establishment' in various countries, at various times: there may be, associated with the universities or the geographical societies, a few dominant figures who form the nucleus of a 'national school'.   There are, however, people whose work is of considerable merit who remain isolated from these schools and whose work is not well known, as in the case of Vallaux in France whose life was spent almost entirely in Brittany, academically a long way from Paris.   Adversity may be a spur to activity, as in the case of Pol in Poland whose work could be regarded as a contribution to national consciousness among a people resenting government by other powers.

Education has been a main concern of many geographers and the great achievement of Cortambert and Levasseur in France bore fruit in the status of the subject in French schools, colleges and universities. Levasseur, at a time when versatility of scholarship was fairly common, was notable for the breadth of his learning in history and economics as well as in geography, though only his historical writing is now regarded as important.   Anuchin, for example, a follower of Darwin, worked on archaeology, anthropology and ethnography as well as geography, and made a fine contribution to scientific education in Russia.   In geography, as in other natural sciences, fieldwork is needed and in Britain Geoffrey Hutchings was fortunate to realize his vision of field study centres for young people under a national scheme which attracted helpful support from university geographers.   However, academic careers are so varied that no standard pattern may be discerned.   One remarkable Russian, Voyeikov, was sufficiently wealthy to travel widely, accumulating material on climatology, which he regarded as the key to physical geography, and became an international authority.   His interest in applied geography developed on a climatological basis and was mainly in crops, irrigation, winter snows, spring thaws, and the varied character of soils.   In the very different circumstances of the United States Brigham began his professional life as a supporter of William Morris Davis but in time became critical of some of his theories, and also of the contemporary correlations made between physical and human phenomena.   He shared the wish of his colleagues for the improvement of geographical education.

What geographers have done from classical times to the present day has naturally varied according to the views they accepted and in some cases vigorously defended.   In all one may recognize a concern with man and the earth.   This concern has taken forms that are many and varied, for both from the physical and human aspects there are several possible interpretations of observed phenomena and controversy has flourished. The ideas discussed in this volume come from the problems confronting people over a wide span of time and range of place.   The searcher for truth may become a specialist looking at problems microscopically --

clearly a necessary condition of progress -- but there are above and beyond such specialisms broad themes of abiding human interest.   Some ideas have been thoughtlessly discarded only to rise up again at a later time, helpful to the emergence of new thought, which is never advanced by discarding all that has gone before, but should rather gather wisdom from the past with discrimination and insight.

T.W. Freeman

Note:

Intending authors are asked to write to Professor T.W. Freeman, c/o Department of Geography, The University, Manchester M13 9PL, who will send a note of information for authors of biobibliographical studies.

List of Abbreviations

Abbreviations have been adopted from *British Standard 4148: Part 2, 1975 Word-abbreviation list*, and refer to abbreviations in both the bibliographical references and chronological tables

Aberdeen Univ. Rev. Aberdeen University Review
Abstr. Philos. Trans. (R. Soc.) Abstracts, Philosophical Transactions (Royal Society)
Actas ... Congr. Int. de Americanistas Actas ... Congress Internacional de Americanistas
Allg. Dtsch. Biog. Allgemeine Deutsche Biographie
Altpreussische Biog. Altpreussische Biographie
Am. J. Sci. American Journal of Science
Am. J. Sociol. American Journal of Sociology
Amani Mem. Amani Memoirs
Ann. Am. Acad. Polit. Soc. Sci. Annals of the American Academy of Political and Social Science
Ann. Assoc. Am. Geogr. Annals of the Association of American Geographers
Ann. Geogr. Annales de Géographie
Ann. Soc. Lett. Sci. et Arts des Alpes-Marit. Annales de la Société des Lettres, Sciences et Arts des Alpes-Maritimes
Annu. Rep. Smithson. Inst. Annual Report of the Smithsonian Institution
Archit. d'Aujourd'hui Architecture d'Aujourd'hui
Archit. Yearb. Architect's Yearbook
Ausl. Ausland
Bibl. Bibliog. Bibliotheca Bibliologica
Bibl. Bibliogr. Bibliothèque Bibliographie
Bombay Geogr. Mag. Bombay Geographical Magazine
Br. Assoc. Advance. Sci. British Association for the Advancement of Science

Br. J. for the Hist. of Sci. British Journal for the History of Science
Bul. Eugenie Buletinul de Eugenie şi Biopolitică
Bul. Soc. (regale) Rom. Geogr. Buletinul Societăţii (regale) Rômane de Geografie
Bull. Am. Geogr. Soc. Bulletin of the American Geographical Society
Bull. Assoc. Géogr. Fr. Bulletin de l'Association des Géographes Françaises
Bull. Geogr. Soc. Chicago Bulletin of the Geographical Society of Chicago
Bull. Geol. Soc. Am. Bulletin of the Geological Society of America
Bull. Inst. Int. Stat. Bulletin de l'Institut Internationale de Statistique
Bull. Inst. Océanogr. Bulletin de l'Institut Océanographique
Bull. Soc. Géogr. Bulletin de la Société de Géographie (Paris)
C.R. Acad. des Sci. Compte-Rendus, Académie des Sciences
C.R. Congr. Int. Géogr. Compte-Rendus, Congrès Internationale de Géographie
C.R. Etud. Géogr. Compte-Rendus, Etudes Géographiques
Cah. Géogr. Cahiers de Géographie
Calcutta Geogr. Rev. Calcutta Geographical Review
Calcutta Rev. Calcutta Review
Contemp. Rev. Contemporary Review
Dumfriesshire and Galloway Nat. Hist. and Antiq. Soc. Trans. J. Dumfriesshire and Galloway Natural History and Antiquarian Society, Transactions and Journal

East Afr. Agric. J. East African Agricultural
Journal
East Afr. Agric. Res. Stn. East Africa Agricultural
Research Station
Econ. Hist. Rev. Economic History Review
Erdkd. Erdkunde
Erdwiss. Forsch. Erdwissenschaftliche Forschung
Et. Classiques Etudes Classiques
Festschr. Festschrift
Geogr. J. Geographical Journal
Geogr. Rev. Geographical Review
Geogr. Rev. India Geographical Review of India
Géogr. Univ. Géographie Universelle
Géogr. Univ. Quillet Géographie Universelle Quillet
Geogr. Z. Geographische Zeitschrift
Geol. Mag. Geological Magazine
Geol. Soc. Am. Bull. Geological Society of America
Bulletin
Geopub (Rev. Geogr. Lit.) Geopub, Review of Geo-
graphical Literature
Heidelberger Geogr. Arb. Heidelberger Geographische
Arbeiten
Hum. Relations Human Relations
I.G.U. Comm. Med. Geogr. International Geo-
graphical Union Commission on Medical
Geography
Illinois Geol. Surv. Bull. Illinois Geological
Survey Bulletin
Indian J. Econ. Indian Journal of Economics
Indian J. Med. Res. Indian Journal of Medical
Research
Indian Sci. Congr. Assoc., Anthropol. Archaeol. Sect.
Indian Science Congress Association,
Anthropological and Archaeological Section
Inst. Br. Geogr. Trans. Pap. Institute of British
Geographers, Transactions and Papers
Int. Encycl. Soc. Sci. International Encyclopaedia
of the Social Sciences
Int. Geogr. Congr. Rep. International Geographical
Congress Report
Int. Mon. International Monthly
Izv. Russkogo Geogr. Obshch. Izvestiya Russkogo
Geograficheskogo Obshchestva
J. Am. Inst. Plann. Journal of the American Institute
of Planning
J. Ecol. Journal of Ecology
J. Econ. Journal des Economistes
J. Farm Econ. Journal of Farm Economics
J. Geogr. Journal of Geography
J. Madras Geogr. Assoc. Journal of the Madras
Geographical Association
J. New York State Teach. Assoc. Journal of the New
York State Teachers Association
J. Proc. R. Aust. Hist. Soc. Journal and Proceed-
ings of the Royal Australian Historical
Society
J. R. Inst. Br. Archit. Journal of the Royal
Institute of British Architects
J. Savants Journal des Savants
J. Soc. Res. Journal of Social Research
J. Town Plann. Inst. Journal of the Town Planning
Institute
Kentucky Geol. Surv. Bull. Kentucky Geological Survey
Bulletin
La Géogr. La Géographie
Mag. of Art Magazine of Art

Mém. Acad. Sci. Morales et Polit. Inst. Fr. Mémoires
de l'Académie des Sciences Morales et Politiques
de l'Institut de France
Mod. Rev. Modern Review
Mus. J. Museums Journal
N. Am. Rev. North American Review
N. Br. Rev. North British Review
N.Z. Geogr. New Zealand Geographer
N.Z. Geol. Surv. Bull. New Zealand Geological Survey
Bulletin
N.Z. J. Geol. and Geophys. New Zealand Journal of
Geology and Geophysics
Nat. Nature
Natl. Acad. Sci. Biog. Mem. National Academy of
Sciences, Biographical Memoirs
Neue Z. für Syst. Theologie und Religionsphilos.
Neue Zeitschrift für systematische Theologie
und Religionsphilosophie
New Mex. Hist. Rev. **New Mexico Historial Review**
New Repub. New Republic
Pac. Northwest Q. Pacific Northwest Quarterly
Paideuma (Mitt. zur Kulturkunde) Paideuma
(Mitteilungen zur Kulturkunde)
Petermanns Geogr. Mitt. Petermanns Geographische
Mitteilungen
Philos. Trans. (R. Soc.) Philosophical Transactions
(Royal Society)
Plann. Outlook Planning Outlook
Pol. Słownik Biogr. Polski Słownik Biograficzny
Prakti. Med. Prakticheskaya Meditsina
Proc. Am. Philol. Assoc. Proceedings of the
American Philological Association
Proc. Am. Philos. Assoc. Proceedings of the American
Philosophical Association
Proc. ... Educ. Conf. Proceedings of the ...
Educational Conference
Proc. Indian Geogr. Soc. Proceedings of the Indian
Geographical Society
Proc. R. Geogr. Soc. Proceedings of the Royal
Geographical Society
Proc. R. Soc. Edinburgh Proceedings of the Royal
Society of Edinburgh
Proc. R. Soc. N.Z. Proceedings of the Royal Society
of New Zealand
Prospect R. Incorporation Archit. Scotland Q.
Prospect, Royal Incorporation of Architects in
Scotland, Quarterly
Q. California Hist. Soc. Quarterly of the California
Historical Society
Rep. Superintendent Coast Surv. Report of the
Superintendent of the Coast Survey
Rev. Adamachi Revista Ştientifică Adamachi
Rev. d'Hist. des Sci. Revue d'Histoire des Sciences
Rev. d'Hist des Sci. Applic. Revue d'Histoire des
Sciences et de leurs Applications
Rev. Deux Mondes Revue des Deux Mondes
Rev. Econ. Int. Revue Economique Internationale
Rev. Gen. Sci. Revue Générale des Sciences Pures
et Appliquées
Rev. Géogr. Revue Géographique
Rev. Geogr. Rom. Revista Geografică Română
Rev. Int. Enseign. Revue Internationale d'Enseignement
Rev. Polit. et Litt. Revue Politique et Littéraire
Rev. Polit. et Parlementaire Revue Politique et
Parlementaire
Rev. Sci. Revue Scientifique

Rev. Sci. Polit. Revue des Sciences Politiques
Sci. Am. Scientific American
Sci. Cult. Science and Culture
Scots Mag. Scots Magazine
Scott. Bookman Scottish Bookman
Scott. Geogr. Mag. Scottish Geographical Magazine
Scott. Stud. Scottish Studies
Séances et Trav. de l'Acad. Sci. Morales et Polit.
    Séances et Travaux de l'Académie des Sciences
    Morales et Politiques (de l'Institut de France)
Smithson. Inst. Bur. Am. Ethnol. Bull. Smithsonian
    Institution Bureau of American Ethnology
    Bulletin
Sociol. Pap. Sociological Papers
Sociol. Rev. Sociological Review
Soil Res. Br. Emp. Soil Research in the British
    Empire
Sudhoffs Arch. Sudhoffs Archiv für Geschichte der
    Medizin und der Naturwissenschaften
Tasmanian J. Nat. Sci. Tasmanian Journal of
    Natural Sciences
Town Plann. Rev. Town Planning Review
Trans. Antropol. Sect. Soc. Nat. Sci. Transactions
    of the Anthropological Section, Society for
    Natural Sciences
Trans. Int. Congr. Soil Sci. Transactions of the
    International Congress of Soil Science
Trans. N.Z. Inst. Transactions of the New Zealand
    Institute
Trans. Oneida Hist. Soc. Transactions of the Oneida
    Historical Society
Trans. Proc. Am. Philol. Assoc. Transactions and
    Proceedings of The American Philological
    Association
Trans. Proc. Geogr. Soc. Pac. Transactions and
    Proceedings of the Geographical Society of
    the Pacific
Trans. R. Soc. N.Z. (Geol.) Transactions of the
    Royal Society of New Zealand (Geology)
Trans. Town Plann. Congr. Transactions of the Town
    Planning Congress
Trav. Inst. Géogr. Univ. Cluj Travaux de l'Institut
    Géographique de l'Université de Cluj
Trav. Sec. Travaux de Section
Trudy Antropol. Otdeleniya Trudy Antropologii
    Otdeleniya
Trudy Inst. Etnog. Trudy Instituta Etnografii
Univ. California Publ. Geogr. University of
    California Publications in Geography
USSR Acad. Sci. Publ. USSR Academy of Sciences
    Publications
Z. für Relig. und Geistesgesch. Zeitschrift für
    Religions- und Geistesgeschichte
Z. Osterreich. Gesell. Meteor. Zeitschrift für
    Osterreichischen Gesellschaft für Meteorologie
Zap. Mineral. Obshch. Zapiski Mineralogicheskogo
    Obshchestva
Zap. po Gidrografii Zapiski po Gidrografii
Zap. Russkogo Geogr. Obshch. po Obshchei Geogr.
    Zapiski Russkogo Geograficheskogo
    Obshchestva SSSR

Dmitry Nikolaevich Anuchin
1843–1923

VASILY ALEXEYEVICH ESAKOV

An eminent Russian geographer, anthropologist, enthnographer and archaeologist, Anuchin was the founder of the Russian school of geography at the University of Moscow, a scholar of remarkable breadth, and a devoted student of the central areas of the Russian plain. He founded the journal *Zemlevedeniye (Physical Geography)* and regarded geography as a complex of the sciences of nature and man.

1. EDUCATION, LIFE AND WORK

D.N. Anuchin was the sixth child of his family. His father, a retired army officer, was made a hereditary nobleman for his service in the Patriotic War against Napoleon in 1812 and worked in the supply department of the Tsar's court. His mother was of peasant stock but well educated with an interest in music and a good command of French. The family were relatively wealthy and lived in a small mansion with a garden in St. Petersburg. Dmitry's elder brothers Mikhail and Alexander were both at Grammar schools when he was born and went later to the School of Military Engineering and to the Law Department of St. Petersburg University respectively.

The home environment was obviously favourable, with a good library and a plentiful supply of periodicals. Dmitry learned to read and write very early and at the age of seven was already composing essays. At school his favourite subjects were science, literature and history and later he said that he 'particularly liked geography and the stories told by geographers'. In fact his schooling was shaping his world outlook and his reading included the works of

Belinsky and Hertzen, Pushkin and Zagoskin, Victor Hugo and Alexandre Dumas, Gogol, Walter Scott and many other eminent writers. As a child he imitated their writing and developed an easy style.

His manuscript memoirs of 1917 show that as a grammar-school pupil he looked forward to a cultivated life for he expected

> To read books in French and German, to learn English and Italian; to read Schiller, Goethe, Shakespeare and Byron; to translate from Goethe's 'Faust' and 'Egmont', from 'Hamlet' and from Schiller; to buy a microscope and make various observations; to go to places where coins and medals are sold; to visit the Public Library and read books; to pay visits to the Hermitage, the Tsar's Palace, the Rumyantsev Museum and Tsarkoe Selo; to read foreign books; to gather flowers and plants; to collect insects; to make a collection of banned verses; to buy a telescope; to recite; to make meteorological observations; to read Shishkov, Derzhavin, Aksakov, Griboyedov, Polevoi ... and others.

Both his parents died in 1857 and Dmitry was put in the care of his elder brother Mikhail. In 1860 he entered the History and Philology Faculty of St. Petersburg University but his studies were interrupted by tuberculosis and he spent more than two years, 1860-3, in Germany, Italy and France. He used this time profitably for what he termed his 'self-upbringing', which included the study of Italian, Latin and English. (He already knew French

and German.)   He was attracted by the progressive
ideas circulating among Russian revolutionary demo-
crats eager to see changes and maintained his interest
in the progress of Russian science.   And with this he
developed an increasing knowledge of the natural
sciences and philosophy.   On his return home in July
1863 he entered the Natural Sciences Department of the
Faculty of Physics and Mathematics at Moscow University.
All his later life and work was to be centred in Moscow.
   Although Darwin's *Origin of species* was published
only in 1859, the ideas of evolution and of the interaction of all natural phenomena were enthusiastically
taught in Moscow university by a number of Professors,
including S.A. Usov and A.P. Bogdanov, who were pupils
of K.F. Rulyo.   Anuchin became interested in anthropology and ethnography and after graduating in 1867 he
became the Academic Secretary of the Animals and Plants
Acclimatization Society.   His work on anthropology,
archaeology and ethnography continued through the 1870s
and he was invited to be head of a new Department of
Anthropology.   In 1876 he was sent for two years to
visit Paris, London, Vienna and other university towns
to study anthropological museums, attend lectures by
leading scientists and work in their laboratories;   and
in 1878 he organized the Russian section at the World
Anthropological Exhibition in Paris.   French scientists, including Broca and Topinar, showed that Russian
anthropologists were doing creative research and had
developed a new method of craniological research.   In
1879 Anuchin attended the Sixth Congress of Russian
Natural Scientists and Physicians in St. Petersburg.
He also became a member of the Russian Geographical
Society and the diploma announcing this was given to
him by P.P. Semenov - Tian-Shansky.
   Anthropology was now Anuchin's main concern and in
the academic year 1879-80 he gave the first university
course ever provided in Russia on physical anthropology,
inaugurated with a lecture on its scope and aims and
also its links with the natural sciences and history.
His second (Master's) degree was awarded in 1881 for a
thesis, 'On some anomalies of the human skull with particular reference to their distribution according to
race'.   During the next twenty years Anuchin was immensely active as a teacher and as a writer on archaeology, anthropology, ethnography and geography.   He
became head of the new Department of Geography and
Ethnography where he was responsible for the first
university courses ever given in Russian on physical
and regional geography.   In 1889 his research on the
geographical distribution of the male population of
Russia according to height was published.   For this
work Moscow University gave him the D.Sc. degree
(honoris causa) and the Russian Geographical Society
awarded a gold medal.
   In 1890 Anuchin became president of the Society
for the Natural Sciences section on Anthropology and
Ethnography which was supported by the few people
friendly to geography and anthropology then living in
Moscow.   Anuchin now began his detailed field researches in central Russia and especially in the vast
areas around the upper reaches of the Dnieper, western
Dvina and Volga rivers.   Several expeditions were
organized as well as his individual field work, and
the publications on these researches were well-
received in Russia and other countries.   In this
work, as in all he did, Anuchin showed his talent for

organization.   This was notably seen in his work for
the International Congress on Prehistoric Archaeology,
Anthropology and Zoology held in Moscow in 1892 and
especially in his arrangement of a geographical exhibition, which was a fine display of material on the
physical geography of Russian with adjacent areas of
Asia and western Europe.
   Meanwhile the Geographical section, founded by
Anuchin, of the Society for the Natural Sciences,
was gaining strength and the publication of the
journal *Zemlevedeniye (Physical Geography)* from 1894
made Moscow second only to St. Petersburg as a
centre of scientific geography in Russia.   Among
its many distinguished geographers were L.S. Berg,
A.A. Borzov, A.A. Kruber, A.S. Barkov, M.S. Bodnarsky,
I.S. Shchukin and B.F. Dobrynin.
   Anuchin's fruitful work in Moscow University,
his vigorous activity in the Society for the Natural
Sciences, his widespread research of significance to
zoology, anthropology and ethnography as well as to
geography, all contributed to his election as a full
member of the Academy of Sciences in St. Petersburg
in 1896 and as an honorary member in 1898.   Though he
worked in Moscow to the end of his life, Anuchin kept
in touch with the Academy and was particularly interested in their discussions on Russia's potential
resources, and their study of the Arctic regions in
the years immediately preceding the October revolution
of 1917.   His view was that a special Institute of
Geography should be attached to the Academy for wide-
ranging research on these problems.
   During his last years Anuchin remained active and
well able to adjust himself to changed living conditions.   He was concerned with various government
enterprises, particularly with the State Planning
Department and on the recommendation of V.I. Lenin was
commissioned to compile and edit the first *Soviet
school atlas of the world*.   He died on 4 July 1923
and was buried at the Vagankov cemetery in Moscow.
Twenty-five years later the Soviet government immor-
talized his name by special decree.   Selected works
have been published;   memorial plaques have been hung
on walls in Moscow University and at his home,
6a Khlebny Pereulok, and a D.N. Anuchin prize for
the best scientific work on geography has been
established with Anuchin scholarships for students
at Moscow University.

2. *SCIENTIFIC IDEAS AND GEOGRAPHICAL THOUGHT*

Anuchin was a materialist convinced by the theories of
Charles Darwin on evolution, and based all his work on
this methodological foundation.   He viewed geography
from a wide scientific background and first explained
his outlook in *The history of physical geography* in
1885.   Already by that time he thought that geography
was an independent field of knowledge lying between
the natural sciences and the humanities: so too was
anthropology.   He made a distinction between general
and regional geography.
   Physical geography was divisible into three
parts: first inorganic nature, including meteorology
and climatology, hydrography and oceanography,
orography;   second, organic nature (biogeography),
discussing plants and animals;   third anthropogeography, to use the term of his time instead of the

later 'human geography'. This last he did not sub-
divide in 1885 for his early scientific training led
him primarily to the physical aspects of the subject
and the interest in human activity developed with
time, notably with his growing interest in regional
geography. Regional or 'special' geography was the
second major division of the subject in his view.

Humboldt and Ritter were dominant figures of
international significance in the first half of the
nineteenth century and Anuchin noted that Ritter's influence was far greater than Humboldt's. Like many
other geographers, however, Anuchin regarded Humboldt
as the finer scientist for he could not agree with
Ritter's theological views. Anuchin's teaching in
his lectures on physical and regional geography and
in the field work he shared with his students in the
various regions of Russia followed the pragmatic
approach of Humboldt, given fresh impetus in 1859, the
year both of Charles Darwin's *Origin of species* and of
the death of Humboldt and Ritter. In later essays of
1892, 1902, 1912 and 1915 Anuchin showed that he still
accepted his earlier views but of more recent geographers he was a particular admirer of Richthofen.

Anuchin opposed the views of the German geographer A. Hettner, who early in the twentieth century
regarded geography as a chorographic science 'on the
expanses of the Earth's surface according to their
material content'. Hettner, as Anuchin understood
him, believed that the study of the essence of objects
and phenomena, like questions of development, was alien
to geographical science. Anuchin's mind was on the
constant changes in the earth's surface and he argued
that 'an adequate understanding of a country's surface
forms, its landscapes and life can only be attained by
examining its past and studying those processes that
have led to subsequent changes' (*Selected Geographical
Works*, 1949, p.34). And also he argued in 1902:

> The purpose of geography has always been the
> same: our planet, the Earth, in its relation-
> ship to other celestial bodies and -- even
> more important -- in itself, especially its
> surface, serves as the arena for the activity
> of various cosmic and terrestrial forces, as
> a result of which there developed the atmos-
> phere, hydrosphere, lithosphere and pedio-
> sphere even -- if it may be so expressed --
> its biosphere and anthroposphere, that is
> the forms of organic life and the stages and
> forms of culture of its most perfect organic
> life -- man. (Ibid, pp.291-2.)

Therefore Anuchin saw geography as the study of
the surface of the earth, with its inorganic nature
and organic life -- including man -- and the interrelations between them. It was the geographer's
task to analyse the phenomena occurring on the earth's
surface, to compare the changes seen in them, to
classify them, to elucidate their origins and explain
the connections between them. Man is so constantly
in contact with nature that the 'human element' can
never be excluded from geographical study for human
activity is reflected in many distinct features of
the landscape. Geography uses methods similar to
those of other natural sciences, and is based on them,
that is on geology, geophysics, meteorology, botany,
zoology and others.

The established distinction between general and
special or regional geography provided a basis for its
future development in Anuchin's view. He attached
great importance to cartography, though he spoke of it
as an isolated science closely linked to geography;
'the extent of geographical knowledge of a country',
he noted, 'is determined by the extent to which maps
of it have been perfected' (*Geographical works* (1954),
p.314). He wished to see a comprehensive geographical survey of Russia carried out by the Academy of
Sciences, using both the general and the regional
approach. In fact many of his ideas have been
followed by modern Soviet geographers whose research
has dealt with current problems of science and the
economy, and especially with the preservation, pur-
poseful exploitation and transformation of nature.

3. INFLUENCE AND SPREAD OF IDEAS

Anuchin's main works on general geography were
significant in the development of geomorphology and
hydrography. Among them were *The relief of the
surface of European Russia (Relief poverkhnosti
Yevropeiskoi Rossii)* 1895, *Dry Land (Susha)* 1895,
*The lakes of the Upper Volga and the upper reaches of
the Western Dvina (Verkhnevolzhskije ozera i verk-
hoviya Zapadnoi Dviny)* 1897. Anuchin proceeded from
the fact that the contemporary relief of the earth is
the result of the struggle between endogenous and
exogenous forces over a long period, though he thought
that the internal endogenous forces were decisive.
The internal energy of the earth was seen 'in the
compression of strata and particularly in the radio-
active substances which are capable of giving off
heat' (*The origin of man*, 1912, p.33). Nevertheless exogenous forces were significant. Anuchin
also gave a comparative description of various levels
of land relief and discerned three main relief types,
mountain, hill and plain, and for each he showed the
genetically distinct forms of the surface.

Lakes were a source of particular interest to
Anuchin and in his researches on them he dealt with
them not as autonomous entities but in relation to
their geological and geographical setting. He is
justly recognized as one of the founders of limnology
in Russia and many of his students worked on lakes,
notably L.S. Berg, whose work *The Aral Sea (Aralskoye
Morye)* appeared in 1908. Anuchin stressed the importance of systematic meter gauge observation and
also regular study of the snow cover.

Anthropological investigations in Russia were
stimulated by Anuchin who accepted the progressive
ideas of A.P. Bogdanov and other scientists. He
encouraged the growth of the anthropological section
of the Society for the Natural Sciences, especially
on the physical composition of the Russian population,
and himself wrote studies of the Ains, 1876, on the
Great Russians, the Little Russians and other physical
types including those found in northern Asia.

Anuchin regarded anthropology as a broad subject
dealing not only with man's physical nature but also
with his life and activity in everyday circumstances,
both now and in the past, including especially the
prehistoric era. This led him to write *The origin
of man (Proiskhozhdeniye cheloveka)* 1912, in which he
showed that 'The human race is strictly one type ...
the whole of mankind originated from the same fore-

fathers, whose descendants only gradually formed races'.   He condemned racialism and it is largely due to Anuchin that Russian anthropology never became racialist.   Even now his pronouncements are quoted against racialist views.

Ethnography and archaeology were of vast interest to Anuchin and he served these subjects both as a researcher and as a scientific organizer, notably by encouraging their teaching in universities.   His own researches include such works as *Sleighs, boats and horses as appurtenances of the funeral rite (Sani, ladiya i koni, kak prinadlezhnosti pokhoronnogo obryada)* 1890, *On the history of acquaintance with Siberia before Yermak (K istorii oznakomleniya s Sibiryu do Yermaka)* 1890 and many more regarded by Soviet scholars as examples of creative scientific work.

Anuchin worked assiduously on the history of science and was especially concerned with its methodology.   He wrote about the development of sciences in general, and particularly of geography, and about the contribution of individuals (including geographers, anthropologists, travellers and others) to the history of science.   His studies of M.V. Lomonosov, Charles Darwin, Alexander Humboldt, N.N. Miklukho-Maklai and others are of particular value.

One of Anuchin's many achievements that brought enlightenment to young people was to improve science teaching in secondary schools, especially of geography. In fact his activity was so varied -- though inter-related -- that it has been said that he did as much for science in Russia as several institutes.   His pedagogic, scientific and administrative activities have been extensively treated in Soviet literature. To this day scientists return to his creative work for Soviet geographers, anthropologists, ethnographers and archaeologists regard him as a remarkably versatile scientist who gave selfless service to the old Russia in which most of his life was spent and the Soviet Russia whose first few years he lived to see.

Bibliography and Sources

1. OBITUARIES AND REFERENCES ON D.N. ANUCHIN
Sbornik v. chest semidesyatiletiya Professora
    D.N. Anuchina (Papers in honour of the seventieth
    birthday of Professor D.N. Anuchin), Moscow
    (1913), 604p.
Bogdanov, V.V., *D.N. Anuchin Antropolog i Geograf*
    (D.N. Anuchin, anthropologist and geographer),
    Moscow (1941) 66p.
'Pamyati D.N. Anuchina (1843-1923) In memory of
    D.N. Anuchin (1843-1923).'
Trudy Inst. Etnogr. N.N. Miklukho-Maklai (Trans.
    N.N. Miklukho-Maklai Inst. Ethnog.), new ser.,
    vol 1, *USSR Acad. Sci. Publ.* (1947) 282p.
'D.N. Anuchin' in Grigoriev, A.A., *Lyudi russkoi nauki*
    (People of Russian science), vol 1, Moscow-Leningrad
    (1948), 599-605 (2nd. ed.), 1962)
Soloviev, A.I., 'D.N. Anuchin, ego osovniye geogra-
    ficheskiye idie i ego rol v razvitii russkoi geogra-
    fi (D.N. Anuchi his main geographical ideas and his
    role in the development of Russian geography)',

Voprosy Geografii (Questions of Geography), Moscow
    (1948) collection 9, 9-28
Esakov, V.A., 'D.N. Anuchin i sozdaniye russkoi univer-
    sitetskoi geograficheskoi shkoly (D.N. Anuchin and
    the creation of Schools of Geography at Russian
    universities)', *USSR Acad. Sci Publ.*, Moscow
    (1955) 182p.
Karpov, G.V., *Put uchenogo (The path of a scientist)*,
    Moscow (1958), 342p.

2. MAIN WORKS OF D.N. ANUCHIN
1874 'Antropomorfniye obeziyany i nizhnie tipy
    chelovechestva (Anthropomorphic apes and lower
    types of man)', *Priroda (Nature)* book 1, 185-280;
    book 3, 220-76, book 4, 81-141
1876 'Materialy dlya antropologii Vostochnoi Azii.
    Plemya Ainov (Materials for the anthropology of
    Eastern Asia.   The tribe of the Ains).'
    *Trudy Antropol. Otdeleniya Obshch, antrop. i
    etnogr.   (Trans. Anthropol. Sect. Soc. Nat. Sci.).*
    vol 20, book 10, 79-204
1879 *Antropologiya, ego zadachi i methody
    (Anthropology, its problems and methods)*,
    Moscow 20p.
1880 'O nekotorykh anomaliyakh chelovecheskogo cherepa
    i preimushchestvenno ob ikh rasprostranenii po
    rasam (On some anomalies of the human skull with
    particular reference to their distribution accord-
    ing to race)', *Trudy Antropol. Otdeleniya Obshch.
    antrop. i etnogr.   (Trans. Anthropol. Sect. Soc.
    Nat. Sci. Anthropol. Ethnog.)*, vol 6, 120p.
1885 *Kurs Lektaii po istorii zemlevedeniya (Course of
    lectures on the history of physical geography)*,
    (lithogr.), Moscow, 286p.
1887 *Kurs lektaii po obshohei geografii (Course of
    lectures on general geography)*, (lithogr. Moscow,
    260p.
1889 *O geograficheskom raspredelenii rosta muzhskovo
    naseleniya Rossii... (On the geographical distri-
    bution of height among the male population of
    Russia...)* St. Petersburg, 184p.
1895 'Relief poverkhnosti Yevropeiskoi Rossii (the
    relief of the surface of European Russia)',
    Zemlevedeniye (Physical Geography), vol 2, book 1,
    77-126; book 4, 65-124. Also published in
    Anuchin, D.N. and Borzov, A.A., *Relief Yevropeis-
    koi chasti SSSR (The relief of the European
    USSR)*, Moscow (1948), 35-147
1897 *Verkhnevolzhskiye ozera i verkhovye Zapadnoi
    Dviny Rekognostsirovka i issledovaniya.   (The
    lakes of the Upper Volga area and the upper
    reaches of the Western Dvina, reconnaissance and
    research)* 1894-5 Moscow, 156p.
1907 *Yaponiya i Yapontsy (Japan and the Japanese)*,
    Moscow, 133p.
1907 *Kurs fizicheskoi geografii (Course in physical
    geography)*, (lithogr.), Moscow
1912 'Proiskhozhdeniye cheloveka i ego istoricheskiye
    predki (The origin of man and his predecessors)',
    in *Itogi Nauki v Teorii i Practike (The results
    of science in theory and practice)*, Moscow,
    vol 6, 691-782
1949 Berg, L.S. (ed.), *Izbranniye geograficheskiye
    raboty (Selected geographical works)*, Moscow,
    388p.

1950 *D.N. Anuchin o Lyudyakh russkoi nauki i kultury*
     (D.N. Anuchin on people of Russian science and
     culture), Moscow, 335p.
1954 Grigoriev, A.A. (ed.), *Geograficheskiye raboty*
     (Geographical works) Moscow, 472p. and bibliogr.
1960 *Lyudi zarubezhnoi nauki i kultury (People of*
     foreign science and culture), 231p.

Vasily Alexeyevich Esakov is Professor of Geography
and Senior Scientific Assistant in the U.S.S.R.
Academy of Sciences Institute of the History of
Natural Sciences and Technology, Moscow.

CHRONOLOGICAL TABLE: DMITRY NIKOLAEVICH ANUCHIN

Dates	Life and career	Activities, travel, fieldwork	Publications	Contemporary events and publications
1843	Born at St. Petersburg			
1845				Organization of the Russian Geographical Society; *Kosmos* (Al.Humboldt)
1854-60	Studied at schools in St. Petersburg			
1859				*Origin of species* (Charles Darwin)
1860	Studied at University of St. Petersburg			Abolition of serfdom in Russia
1860-3		Visited Germany, Italy and France		
1863-7	Studied at University of Moscow			
1871-4	Academic Secretary of the Animals and Plants Acclimatization Society		*Ocherki afrikanskoi fauni. I.Sekretar.* (1873) (Essays on African fauna)	
1875-6	Teacher of geography in a secondary school			
1876			*Materiali dlja antropologii Wostochnoi Azii. Plemja ainow.* (Materials for anthropology of East Asia. The Ain tribe)	
1876-9		Anthropological and ethnographical studies in France, Germany and Great Britain		
1878		Organized Russian department of the International anthropological exhibition in Paris		
1879	Elected a member of the Russian Geographical Society	Organized anthropological exhibition in Moscow. Delivered lectures on anthropology in Moscow		
1880			*O nekotorih anomalijah chelowecheskogo cherepa* ... (On certain anomalies of the human cranium)	
1881	Master of Anthropology degree	Archaeological studies in Dagestan (N.Caucasus)		
1884-1919	Head of the Department of Geography and Ethnology at Moscow University			

Dates	Life and career	Activities, travel, fieldwork	Publications	Contemporary events and publications
1884	Professor extra-ordinary			
1887		Archaeological studies in the Middle Urals		
1889	Doctor of geography *honoris causa*		*O geograficheskom raspredelenii rosta muzhskogo naselenija Rossii ...* (On the geographical distribution of the growth of male population in Russia)	
1890	President of the Society of Natural Sciences, anthropology, ethnography; also President of the geographical Division of the Society	Founded the geographical Division of the Society for Natural Sciences. Chief of the research expedition in the upper reaches of the Dnepr, the West Dvina and Volga. Expedition to the Caucasus		
1891		International Geographical Congress, Berne. Travelled in Switzerland		
1892		International Anthropological Congress, Moscow. Organized a geographical exhibition in Moscow		
1894		Founded the journal, *Zemlevedeniye*		
1894-1923	Editor-in-chief of the journal *Physical Geography (Zemlevedeniye)*			
1894-5		Chief of the research expedition on the upper reaches of the Dnepr, Volga and West Dvina		
1895		VI. International Geographical Congress, London. Travelled through western Europe	*Relev poverhnosti Evropeiscoi Rossii v posledovatelnom rasvitii o nem predstavlenii* (Relief of the surface of European Russia. Evolution of the knowledge of it). *Susha. Kratkie svedenija po orographii.* (The Land. A brief treatment of its orography)	
1896	Academician of the Academy of Sciences, St. Petersburg			

Dates	Life and career	Activities, travel, fieldwork	Publications	Contemporary events and publications
1898	Honorary academician of the Academy of Sciences, St. Petersburg			
1897			*Verkhnevolzhskie ozera i verhovja Zapadnoji Dvini* (The upper Volga lakes and upper reaches of the West Dvina)	
1907			*Japonija i japonci* (Japan and the Japanese). *Kurs fisicheskoi geografii* (Course of physical geography)	
1917				The Great October Revolution
1919				The Department of Geography and Anthropology was divided into two independent units
1922		Organized a Research Institute of Geography at Moscow University		
1922-3	Director of the Research Institute of Geography at Moscow University			
1923	Died in Moscow			

Nirmal Kumar Bose

1901–1972

SITANSHU MOOKERJEE

Born in Calcutta on 22 January 1901, Bose belonged to a well-to-do family though his father died early in life. Having matriculated from the Sagar Dutt High School he entered the Scottish Churches College, Calcutta for the intermediate course in Science and later went to the Presidency College for the Honours degree in Geology with Chemistry and Physics as subsidiary subjects. In all his courses his work was of first class calibre but he was refused admission to the M.Sc. course, available only at the Presidency College, on account of his political activities. Already he was involved in the non-cooperation movement but this was only the beginning of his struggles as a freedom fighter. He had been appointed as a demonstrator in the Geology Department at the Aligarh University in 1922, but as he could not study for the M.Sc. degree there he returned to Calcutta. He went to Puri to study the famous sun temple of Konarak and was fortunate in attracting the attention of Sir Asutosh Mookerjee, Vice-Chancellor of Calcutta University, who arranged that he should join the recently established Anthropology post-graduate classes in Calcutta University. In 1925 he was awarded the M.Sc. degree in Anthropology at the top of the First Class.

For the next few years Bose combined his social studies with organization work for the Indian National Congress. In 1925 his book entitled *Cultural Anthropology*, written with the help of many advisers and friends, won warm praise from Professor A.L. Kroeber (1876-1960), the American anthropologist. In 1930, when the civil disobedience campaign was widespread, he was strongly implicated and served a prison sentence. To this period belong two other major publications, *Konoraker Bibaran*, 1929 (also 1960), and *Canons of Orissan Architecture*, 1932. The former was written in Bengali and the latter, in English, was of a distinctly technical character. Bose was also a prolific contributor to journals of a literary and scientific nature.

In the 1930s he went to Wardha for discussions with Mahatma Gandhi on the problems of the time. These included the working out of the philosophy of 'Hind Swaraj' which through the years developed into the Gandhian philosophy. Ethically and pragmatically Bose became a socialist of the Gandhian type and he wrote two books on his views, *Studies in Gandhism* 1940 (also 1947 and 1962) and *Selections from Gandhi* 1939 (also 1948, 1957, 1968, 1972). Through this period of his life he was in effect a freelance worker and he was undecided about the offer of an assistant lectureship in Anthropolgy in 1937. During the Second World War he supported the 'Quit India' movement against the British raj and his activities included the editing of the Bengali version of the *Harijan*, the purpose of which was to acquaint the masses with the thought of Gandhi.

In 1942 Gandhi and numerous other political personalities were arrested under the Defence of India laws and Bose was sent to the Dum Dum gaol with other political prisoners from all parts of Bengal. There he reflected on his own extensive travels and collected information from his fellow prisoners to produce a treatise in Bengali on Bengal prior to partition.

This included detailed information, lucidly presented, on soils, hydrology, agriculture, cottage and other small industries and their socio-economic aspects.

Release came in 1945 and Bose then joined the Department of Geography at Calcutta University as Lecturer in Human and Cultural Geography. He was promoted shortly afterwards to the first Readership in Geography at Calcutta University. His concern for Gandhi remained as great as ever and he acted as his personal secretary during a long visit to Bengal. He also accompanied Gandhi on various missions, including the walking tour in Noakhali and Patna from November 1946 to March 1947 which was of considerable historical significance. During the journey through Noakhali he was Gandhi's companion. These missions, however, were interludes in his university career and ended with the assassination of Gandhi on 30 January 1948. Despite his long association with Gandhi, Bose was not a major political figure but rather a worker concerned to encourage constructive progress among the poorest elements in Indian society. During this time he remained active as a researcher and writer and published a number of papers in the *Calcutta Geographical Review* (founded 1936) which became the *Geographical Review of India* in 1950. Bose was editor of this journal to 1954 and again from 1956-8. He also edited the anthropological journal, *Man in India*, from 1951 to his death, and from 1959-64 was Director of the Anthropological Survey of India and Adviser on Tribal Affairs.

Having acquired increasing fame in his dual capacity as a man of affairs and a man of letters, in 1959 Bose was appointed Director of the Anthropological Survey of India. In that post he completely restructured the work of the Survey; his particular achievement was to identify the regional cultural traits as a basis for the eventual social integration of India. A substantial number of reports were produced. Even after his retirement in 1964 the Government of India used his services, notably by engaging him as Commissioner for Scheduled Castes and Tribes from 1967 to 1970. There too he produced a number of reports, many of them with the help of young workers who submitted their work for the Ph.D. degree. Two of the most appreciated works of Bose were his study of the socio-economic character of Calcutta, *Calcutta*, 1964 and *Calcutta: a social survey*, 1968 in which -- as in his other work -- geographical and anthropological aspects were finely interwoven.

Bose was invited to lecture in several universities of the U.S.A. (including Chicago and California), Great Britain and Japan. In 1966 he was awarded the 'Padamashree' (a coveted honour given for distinguished public service) by the President of India. He served as an expert on many committees and was the first geographer to be elected as a fellow of the National Institute of Sciences. He continued to write in Bengali and his *Paribrajaker Diary* and *Nabin-O-Prachin* are admired in his homeland. He was a man of generous spirit giving to numerous charities, mainly in secret. Much revered by a wide circle of friends and admirers, he died in Calcutta on 15 October 1972. He was a man of great ability and conscience whose career shows a fascinating adjustment to the troubled circumstances of his time.

Bose was a bachelor in whom a vigorous and resolute disposition was combined with a marked capacity for the enjoyment of life. He always said that his first love was science, but his conception of science was singularly broad and essentially human for his aim was to understand the mysteries of social change, to find out the causes of human misery and to alleviate them, and to discern the whole magnificent range of human culture. He was a friend of many famous and influential people of his time, including Gandhi and Tagore, and also (even more closely) the Pathan leader of the Northwest frontier, Khan Abdhul Ghaffar Khan. His association with Gandhi was more than a friendship based on temperamental affinity for he thought deeply about the scientific application of the Gandhian philosophy in Indian life. Socially, he was equally at home with his academic peers and with much humbler people with whom he was frequently to be seen enjoying conversation at one of Calcutta's old bookshops. In him there was something of the social historian and he often spoke of himself as just that. His twenty-odd books in English, with almost as many in Bengali, and a hundred articles in journals, show his constant literary activity. At the time of his death he was President of the Asiatic Society and also of the Bangiya Sahitya Parishad (the Bengali Literary Council) and during his last days in a nursing home he was correcting the proofs of his favourite journal, *Man in India*. On his death there were many fine tributes in the press and in learned journals and his students and friends set up a Bose Memorial Foundation in Varansi. Not the least interesting comment came from the Indian geographer, S.P. Chatterjee, who said that the 38 million tribal people of India had lost their greatest champion.

Bibliography and Sources

As noted in the text, much of Bose's writing was on anthropology and he was author of *Fifty Years of Science in India: Progress of Anthropology and Archaeology*, 1963, a volume written for the Indian Science Congress Association, Calcutta. His work on geography shows his concern with both rural and urban problems. On contemporary problems he has much of interest to say as O.H.K. Spate and A.T.A. Learmonth note (*India and Pakistan*, 3rd ed, (1967), 172) of his paper 'Some problems of urbanization,' *Man in India*, vol 42/4 (1962), 255-62, 'a stimulating critique ... combining elements from the thinking of Gandhi and Patrick Geddes'. Notable also is his work on Calcutta, to which reference is made in the text. Obituaries include those of K. Bagchi in *Geogr. Rev. India* vol 34 (1972), 404-6.; B.M. Thiranaranan in *Indian Geogr. J.* vol 47/3-4 (1972), 36-9; and S.L. Kayastha in *Nat. Geogr. J. India* vol 18 (1972), 267-8.

The papers of Bose were so numerous that here only a few, of a geographical character, can be mentioned.

1938 'Certain types of Indian people', *Calcutta Geogr. Rev.*, vol 2/1, 57-61
1939 (with K. Bagchi) 'Geography of the Seraikella state', *Calcutta Geogr. Rev.*, vol 2/2, 35-43
1947 'Bengal Partition and after', *Calcutta Geogr. Rev.*, vol 9, 14-23
1949 'The interview in human geography', *Calcutta Geogr. Rev.*, vol 11/1, 12-15
1949 'Adaptation to environment,' *Calcutta Geogr. Rev.*, vol 11/2, 12-15
1949 'The disintegration of tribal cultures in India', *Calcutta Geogr. Rev.*, vol 11/3-4, 44-7
1950 'Notes on planning field investigation', *Calcutta Geogr. Rev.*, vol 12/3, 32-8
1953 'The role of social sciences in Community Development', *Geogr. Rev. India*, vol 15/1, 1-5
1953 'Tribal welfare', *Geogr. Rev. India*, vol 15/3, 1-5
1955 'Racism and geography', *Geogr. Rev. India*, vol 17/3, 1-5
1956 'Cultural zones of India', *Geogr. Rev. India*, vol 18/4, 1-12
1958 'Social and cultural life of Calcutta', *Geogr. Rev. India*, vol 20 1-46
1961 *Peasant life in India: a study in Indian unity and diversity* (ed. N.K. Bose), Calcutta, 61p.
1965 'Calcutta; a premature metropolis', *Sci. Am.*, no 213, 90-102

Sitanshu Mookerjee is Principal of Morris College and Professor of Geography at the University of Nagpur and President of the Indian Institute of Geography.

Albert Perry Brigham
1855-1932

PRESTON E. JAMES

Albert Perry Brigham was a disciple of William Morris
Davis and, during the years when Davis was outlining
the content and method of geography as a field of
scholarship, no other person did more than Brigham to
clarify and extend the formula Davis had proposed.
In 1904, when the Association of American Geographers
was founded, Brigham was one of the forty-eight original members; for the first decade he was chiefly
responsible for the management of its professional
affairs.   He was named secretary-treasurer in 1904
and continued to act as treasurer until 1907 and as
secretary until 1913.   In 1914 he was elected President.   Brigham was a gifted lecturer and a writer of
elegant prose.   It was he who initiated the critical
discussion of Davis's ideas and thereby helped to set
the tone of geographical thought.

1. EDUCATION, LIFE AND WORK
Albert Perry Brigham was born on a farm near the
village of Perry in up-state New York on 12 June 1855.
The keys to his character are to be found not only in
the habits of hard work developed on a farm in a rural
community just emerging from frontier self-sufficiency,
but also in his resolution of the conflicting attitudes
on life of his father and mother.   Horace A. Brigham,
his father, was a perfectionist.   He was a Baptist who
believed that God should be feared and would inflict
punishment on those who failed to obey the laws of
righteous behaviour.   His father often expressed the
wish that his only son should become a minister of the
church.   But his mother was very different.   Julia
Perry Brigham was less concerned with perfection than

with human compassion and with the creation and enjoyment of beauty.   She believed in a loving God who
could be depended on for help in time of need.   She
died when Albert Perry was only fifteen but her influence remained throughout his life.   Although she
never travelled more than forty miles from her home,
she made the wide world exciting and challenging
to her son.   She provided him with a geography book
(S. Augustus Mitchell, *A System of Modern Geography
adapted to the Capacity of Youth*, 1849) containing 197
pictures which the boy studied for hours, so that he
remembered them many years later when he rediscovered
the book.   He was fascinated by maps.   Both the
father and mother read frequently from the Bible which
had a lasting impression on Brigham's style of writing.
    In 1875, at the age of twenty, Brigham entered
Madison University in Hamilton, New York (now Colgate
University).   Madison offered a liberal arts programme intended to prepare students for the Baptist
Theological Seminary in Hamilton.   In college Brigham
was a brilliant student, excelling especially in both
writing and speaking.   But in his junior year he took
a course in Natural History offered by Professor
Walter R. Brooks (1821-88).   The face of the earth
and the living plants and animals on it were portrayed
as changing and developing rather than static and
everlasting.
    In accordance with his father's wishes, Brigham
spent three years in the Theological Seminary at
Hamilton.   He graduated and was ordained to the
ministry in 1882.   In that same year he married Flora
Winegar of Amsterdam, New York; and he went to his
first pastorate in the Baptist church in Stillwater

(on the Hudson River north of Albany). From 1885 to 1891 he was the minister of a church in Utica, New York. We catch only glimpses of the struggle that must have been going on in the mind of the young minister between the traditional Calvinist Christianity with its emphasis on punishment for sin, and the gentle humanity he had inherited from his mother.

During this period the seed that had been planted by Professor Brooks began to grow. Brigham even remarked, sometime later, that 'no man should be let to escape to the pulpit until he had had a course in historical geology'. During the summers he went out to study the rocks and landforms of central New York. Working by himself, with no further training beyond the elementary geology taught by Professor Brooks, he completed a survey of the area around Utica which was published in 1888 ('The geology of Oneida County'). During the summer of 1889 he attended two field courses offered by Harvard: one, led by N.S. Shaler of Harvard and H.S. Williams of Cornell, studied the landforms of the Genesee Valley south of Rochester; the other, led by W.M. Davis of Harvard, studied the Helderberg Escarpment south of Albany and the Triassic formations of the Connecticut Valley. These experiences convinced him that he would be of greater service as a teacher, studying and demonstrating the 'adaptability of the earth to the needs of man' than he could ever be as a minister.

In 1891 he resigned from the ministry and went to Harvard for advanced study with Shaler and Davis. He completed the Master of Arts degree in 1892 and was immediately appointed to teach geology at Colgate, where he remained until his retirement in 1925. Brigham died in Washington on 31 March 1932.

2. SCIENTIFIC IDEAS AND GEOGRAPHICAL THOUGHT

Brigham's career as a scholar and teacher was focussed on the border zone between geology and geography. He was never concerned to identify a boundary to separate these fields and in fact usually assumed that what he studied and taught might all be included in geology. 'Geology', he said, 'deals with the earth in its coming to be; geography takes up the earth as it is, particularly in relation to man; but no boundary can be drawn between the two subjects'. 'The essential content of geography', he wrote in 1924, 'is so large that the successful cultivation of the whole demands the energies of many experts' ('The Association of American Geographers 1903-23', *Ann. Assoc. Am. Geogr.*, vol 14 (1924), 110).

In his own contributions, whether they were concerned with physical processes or sequences of settlement, he insisted on the necessity of seeking to identify the causes of things and searching for cause and effect sequences. This theme had made Professor Brooks's teaching stimulating both in the classroom and out-of-doors; in the late nineteenth century scholars still believed that proper scientific procedure could lead to the identification of such sequences, so that observed processes could be traced back to specific causes. Brigham's teaching was exciting to students because he insisted that merely to describe some feature of the earth's surface was incomplete until its cause had been discovered.

When Brigham went to Colgate in 1892 he immedi-

ately started writing and publishing papers in which he set forth in his own clear style the ideas he had picked up during his year of graduate study at Harvard. His paper on 'Rivers and the evolution of geographic forms' (1892) was derived directly from Davis's theoretical model of landform sequences, to which Brigham added examples based on his own field observations in New York. The paper on 'The Finger Lakes of New York' (1893) is a fine example of the use of multiple hypotheses -- a method that had been urged by G.K. Gilbert, T.C. Chamberlin and W.M. Davis. By what processes had these unusual lakes been formed? Why did the lakes become narrower and also deeper toward their southern ends? Brigham suggested several possible answers to these questions and discarded one after another until he had marshalled the evidence in support of the process of glacial gouging. 'The deepest lakes', he concluded, 'were made by the heaviest and most persistent ice streams'. His paper on 'The composite origin of topographic forms' (1895) again applied the concepts and methods he had learned from Davis to an enlarged understanding of the features of up-state New York. In this paper he wrote: 'The teacher of physiography has no greater reward than is his when a student assures him that henceforth his native state will be to him a new country, or that he shall see the hills and valleys of his old home with new eyes...' Indeed, his native state, which he already knew and loved, did come alive with new meaning.

These contributions to the physical geography of up-state New York led quite naturally to the production of text books for the use of his students. In 1901 he published a *Textbook of geology* and in 1902 *Introduction to physical geography* which he wrote with Grove Karl Gilbert. The teacher's guide to the geology text provided directions for field trips that could be taken in the vicinity of eighteen of the larger cities of the United States. In his physical geography he sounded a new theme which for a long time had been neglected by students of earth features. He wrote: 'As the earth has had its influence on man, so man has stamped the earth everywhere with his presence'. This new note, which appeared in 1902, marked the shift of Brigham's attention to the 'organic side of geography'. He observed that among the disciples of Davis who were learning to provide 'explanatory descriptions' of the physical characteristics of areas followed by references to so-called 'organic responses', there was a tendency to forget that responses also had to be proved. Brigham insisted that the need for any scientist to seek the causes of things, and to identify sequences of cause and effect, must apply to the organic side of geography as well as the physical side. This led him to search through the record of human history.

The year 1903 was a remarkable one in the development of American geography. In that year two notable books were published, each representing the connections between the physical features of North America and the course of American history since the beginning of European colonization. One of the books, written by a trained historian, was Ellen Churchill Semple's *American history and its geographic conditions*. The other, written by a trained geologist, was Brigham's *Geographic influences in American history*. Semple's book paid slight attention to the physical origin of so-called geographic conditions. Brigham's book placed heavy emphasis on the physical processes but was relatively light on history -- as historians were not slow to point out. Although the emphasis in the two books was different, the points of view of Semple and Brigham were very much alike.

During the decade that followed his book on American history Brigham contributed numerous studies in historical geography. His 'Geography of the Louisiana Purchase' (1904) was a brief summary for the use of teachers. But his several works dealing with routes across the Appalachians were based on detailed map study and field observation. These included 'Good roads in the United States' (1904) and, a year later, 'The great roads across the Appalachians' (1905). His studies of these routes in eastern United States were summarized in *From trail to railway through the Appalachians* (1907). He was also at work on studies of historical changes in the distribution of population in the United States (published in *Geogr. J.* 1908).

Brigham became increasingly concerned during these years by the lack of careful scholarship in the study of 'geographic influences'. Many of the geographers in that period had been trained by geologists and were not in close touch with the writings of anthropologists or other social scientists. In 1907 and again in 1909 Brigham arranged special round-table discussions at meetings of the Association of American Geographers to exchange ideas about procedures for identifying 'influences' and especially for proving them. In 1907, at the Chicago meeting, he invited historians and anthropologists to attend the round-table. In 1909 the round-table discussed 'The organic side of geography, its nature and limits' (published 1910). One can easily see in these events an effort on Brigham's part to provide the balance in the study of geographical questions which Davis urged but was unable himself to provide.

In 1914, when Brigham was President of the Association, his presidential address was on 'Problems of geographic influence'. He followed Davis in specifying that it was the geographer's task to provide a careful scientific description of the physical environment, but that geographers should use caution and common sense in asserting the existence of influences, and that every possible test should be made to ascertain the validity of any general principles that were suggested. He was especially critical of generalizations concerning the influence of climate that he found in the works of Ward, Huntington and Semple. He observed that 'perhaps there is no subject, unless it be politics, on which men say so much and know so little as about climate'. He was especially disturbed by vague and unproved assertions of climatic influence on racial character, skin colour or man's institutions. The infinitely variable factors of the total environment, he insisted, produce infinitely diverse results on body and mind.

Brigham must be credited with starting the critical analysis of Davis's ideas which led, eventually, to the search for alternatives of the 1920s and 1930s. He began the competitive discussion of geographical ideas and methods without in any way seeking to belittle Davis -- indeed he was one of Davis's strongest supporters and admirers. The discussions were exactly what Davis recognized as an essential next step in

the formulation of a model for geographical study.

In the summer of 1915 Brigham made his first visit to Cape Cod. Here was an unusual region, with distinctive physical characteristics, which offered an unusual opportunity to translate general ideas about influences and the procedures for studying them into a specific application of these ideas in a particular place. The book he wrote, and published in 1920, was entitled *Cape Cod and the Old Colony*. In his preface to the published work Brigham explained that the book was 'not a history and... not a geography': though it dealt with physical features the real aim was to show 'the way men have used these lands and waters and come under their influence'. This book is Brigham's greatest work, not least because he had chosen an almost perfect place to demonstrate the ideas he had been formulating concerning an acceptable model of geographic study and interpretation. It is interesting that he expected the book to appeal not only to a narrow circle of professional colleagues but also to a wider audience of non-specialists.

3. *INFLUENCE AND SPREAD OF IDEAS*

Brigham's influence on the development of geography in the United States was pervasive but not always obvious. He was a quiet worker whose attention to administrative detail was of critical importance in translating Davis's vision of a new professional field into a reality. That Brigham was able to manage the affairs of the profession for many years without arousing antagonism is a measure of the confidence of his colleagues in his sincerity and good judgement. He did much to make American geographical ideas known to Europeans, especially in England and Germany. Throughout his life he was concerned with the improvement of the teaching of geography at all levels.

Even while still in the ministry he was interested in the education of young people. In his own experience the elementary teaching of geography had lacked contact with the real world outside the classroom. Therefore when Davis attempted to reform the teaching of general science in the schools by giving the programme a core of geography, Brigham gave the movement his enthusiastic support. In 1892 and again in 1899 the National Education Association had supported programmes to improve the position of geography. In 1909 Brigham wrote optimistically that the report prepared in 1899 had 'been effective in directing and stimulating the progress of rational geography in the United States'. But a few years later he was shocked by what he found still being taught in the secondary schools. He found, as he described it, 'bits and shreds of geography hanging in the social studies and in general science curricula, where the teachers were so poorly prepared in geography that they could do no more than require the memorizing of meaningless information'. In the 1920s he published several papers on the teaching of geography. But in this field Brigham was frustrated, as most other geographers have been, in the effort to insist that only adequately trained teachers should be assigned to teach geography.

Brigham's influence as a 'geographic envoy' involved appearances at several International Geographical Congresses, and courses of lectures in universities, at home and abroad. His first appearance before an International Congress was in September 1904 in Washington, where he presented a paper on 'Geography and history in the United States'. His first trip to Europe was in 1907, when he went to Switzerland for a rest. But he was back again the next year for the Ninth International Geographical Congress at Geneva. The paper he presented then was on 'The changing patterns of population in the United States'. He lectured at Oxford on the same subject. In 1913 he was granted a year's leave of absence from Colgate and he spent the time living in Germany and visiting other parts of Europe. He left Europe in 1914 just before the outbreak of the First World War.

After the war, and before his retirement in 1925, he was much sought after as a lecturer. He taught in summer school courses at Harvard, Cornell and Wisconsin; and in 1923 and 1925 he lectured at Oxford and at the Royal Geographical Society, London. In the aftermath of the First World War the message that he brought to the British underlined the need to re-establish confidence in the world-wide brotherhood of man. The growing isolationism in the United States he thought was a serious mistake. In London he spoke on the raw-material dependency of the United States, which he presented as part of the interdependence of all mankind (*Geogr. J.*, vol 47 (1924), 412-25).

Brigham continued to lead an active professional life after his retirement from Colgate. He was appointed to two positions in Washington: he was named as Honorary Science Adviser to the Library of Congress; and Vice-Chairman of the Division of Geology and Geography of the National Research Council. These appointments made it necessary to spend the winters in Washington, where he and his wife were soon in the midst of a new circle of friends and professional colleagues.

Brigham received numerous honours which bear witness to his wide influence among professional geographers on both sides of the Atlantic. He received three honorary degrees: Sc.D. from Syracuse University in 1918; Litt.D. from Franklin College in 1920; and LL.D. from Colgate at the time of his retirement in 1925. In 1930 he was named as one of the official delegates of the American Geographical Society to the Centenary of the Royal Geographical Society, London. On the occasion of his seventy-fifth birthday in 1930 the June number of the *Annals of the Association of American Geographers* was filled with 'An appreciation of the contributions to earth sciences of Albert Perry Brigham on his 75th birthday'.

Bibliography and Sources

1. OBITUARIES AND REFERENCES ON A.P. BRIGHAM
Geogr. Rev., vol 22 (1932), 499-500
J. Geogr., vol 31 (1932), 265-6
'An appreciation of the contributions to earth science of Albert Perry Brigham on his 75th birthday', *Ann. Assoc. Am. Geogr.*, vol 20 (1930), 51-104
Keith, Arthur 'Memorial of Albert Perry Brigham', *Bull. Geol. Soc. Am.*, vol 44 (1933), 307-11

Burdick, Alger E. 'The contributions of Albert Perry Brigham to geographic education', unpublished Ph.D. Dissertation at George Peabody College for teachers, 1951

2. SELECTIVE AND THEMATIC BIBLIOGRAPHY OF A.P. BRIGHAM'S WORKS

a. Physical geography
1888 'The geology of Oneida County', *Trans. Oneida Hist. Soc. 1887-1889*, 102-18
1892 'Rivers and the evolution of geographic forms', *Bull. Am. Geogr. Soc.*, vol 24, 23-43
1893 'The Finger Lakes of New York', *Bull. Am. Geogr. Soc.*, vol 25, 203-23
1895 'Drift boulders between the Mohawk and Susquehanna Rivers', *Am. J. Sci.*, 3rd ser, vol 49, 213-28
---- 'The composite origin of topographic forms', *Bull. Am. Geogr. Soc.*, vol 27, 161-73
1897 'Glacial flood deposits in Chenango Valley', *Bull. Geol. Soc. Am.*, vol 8, 17-30
1898 'Topography and glacial deposits of Mohawk Valley', *Bull. Geol. Soc. Am.*, vol 9, 183-210
1906 'The fiords of Norway', *Bull. Am. Geogr. Soc.*, vol 38, 337-48
1929 *The glacial geology and geographic conditions of the Lower Mohawk Valley*, New York State Museum Bulletin, Albany, 280p.

b. Historical and regional geography
1903 *Geographic Influences in American History*, Boston, 366p.
1904 'The geography of the Louisiana Purchase', *J. Geogr.*, vol 3, 243-50
---- 'Good roads in the United States', *Bull. Am. Geogr. Soc.*, vol 36, 721-35
---- 'Geography and history in the United States' in *Rep., 8th Int. Geogr. Congr.*, 958-65
1905 'Great roads across the Appalachians', *Bull. Am. Geogr. Soc.*, vol 37, 321-39
1907 *From trail to railway through the Appalachians*, Boston, 188p.
1920 *Cape Cod and the Old Colony*, New York 284p.
1924 'The Appalachian valley', *Scott. Geogr. Mag.*, vol 40, 218-30
1927 *The United States of America, Studies in physical, regional, industrial and human geography*, London, 308p.

c. Teaching and text books
1901 *Textbook of geology*, New York, 477p.
1902 (with G.K. Gilbert) *Introduction to physical geography*, New York, 380p.
1911 *Commercial geography*, Boston, 489p.
1916 (with C.T. McFarlane) *Essentials of geography*, New York, 266p.
1920 (with C.T. McFarlane) *Results of the World War*, New York, 82p.
---- 'Geography and the War', presidential address, National Council of Geography Teachers, *J. Geogr.*, vol 19, 89-103
1921 'The teaching of geography', *J. New York State Teach. Assoc.*, Feb. 1921, 16-18
---- 'Geographic education in America', *Annu. Rep. Smithson. Inst.*, 487-96

1922 'A quarter century in geography', *J. Geogr.*,
 vol 21, 12-17
1929 'Geography as a cultural factor in education',
 Proc. 9th Educ. Conf., Ohio State University

d. *Philosophical*
1910 'The organic side of geography.   Its nature and
 limits', *Bull. Am. Geogr. Soc.*, vol 42, 442-52
1915 'Problems of geographic influence', presidential
 address, Assoc. Am. Geogr., *Ann. Assoc. Am.
 Geogr.*, vol 5, 3-25
1919 'Principles in the determination of boundaries',
 Geogr. Rev., vol 7, 201-19

e. *Miscellaneous*
1908 'The distribution of population in the United
 States', *Geogr. J.*, vol 32, 380-9
1909 'William Morris Davis of Harvard University',
 Geographenkalendar, Gotha, 1-73
1915 'History of the Transcontinental Excursion', in
 Memorial Volume, Am. Geogr. Soc., 9-45
1922 'Environment in the history of American agriculture', *J. Geogr.*, vol 21, 41-9
1924 'American dependence on foreign products',
 Geogr. J., vol 47, 412-25
---- 'The Association of American Geographers 1903-
 1923', *Ann. Assoc. Am. Geogr.*, vol 14, 109-16

*Preston E. James is Maxwell Professor Emeritus,
Syracuse University, N.Y., U.S.A.*

CHRONOLOGICAL TABLE: ALBERT PERRY BRIGHAM

Dates	Life and career	Activities, travel, fieldwork	Publications	Contemporary events and publications
1855	Born at Perry, New York, 12 June			
1875	Entered Madison University			
1879	Graduated; entered Theological Seminary, Hamilton			
1882	Ordained in ministry; married Flora Winegar; took up pastorate at Baptist Church, Stillwater, N.Y.			
1885	Minister at Utica, N.Y.			
1888			'The geology of Oneida County'	
1889		Attended field courses led by N.S. Shaler, H.S. Williams and W.M. Davis		
1891	Resigned from ministry; entered Harvard			
1892	M.A. degree; appointed to teach geology at Colgate University		'Rivers and the evolution of geographic forms'	National Education Association Committee of Ten
1893			'The Finger Lakes of New York'	
1895			'The composite origin of topographic forms'	
1901			*Textbook of geology*	National Education Association Committee on College Entrance Examinations
1902		Chief interest shifted to human geography	(with G.K. Gilbert) *Introduction to physical geography*	
1903			*Geographic influences in American history*	*American history and its geographic conditions* (E.C. Semple)
1904		8th Int. Geogr. Congr., Washington. Nominated Secretary-Treasurer, Assoc. Am. Geogr. (secretary till 1913; treasurer till 1907)		Association of American Geographers founded
1905			'Great roads across the Appalachians'	

Dates	Life and career	Activities, travel, fieldwork	Publications	Contemporary events and publications
1907		Visited Switzerland	*From trail to railway through the Appalachians*	
1908		9th Int. Geogr. Congr., Geneva: lectured at Oxford University		
1910			'The organic side of geography'	
1912		Am. Geogr. Soc. Trans-Continental Excursion		
1913		Visited Berlin and Oxford		
1914		President, Assoc. Am. Geogr.		Outbreak of war
1915			'Problems of geographic influence' (Presidential Address)	
1918	Sc.D. Syracuse University			
1919			'Principles in the determination of boundaries'	
1920	Litt.D. Franklin College		*Cape Cod and the Old Colony*	
1923		Lectured at Oxford		
1924			'American dependence on foreign products'	
1925	Retired from Colgate University; LL.D. Colgate University	Lectured at Oxford and at R. Geogr. Soc., London		
1927		Science adviser, Library of Congress	*The United States of America*	
1929		Vice-Chairman, Division of Geology and Geography, National Research Council	'The glacial geology and geographic conditions of the Lower Mohawk Valley'	
1930		Am. Geogr. Soc. delegate to R. Geogr. Soc. centenary: lectured in Havana, Cuba		
1932	Died at Washington, 31 March			

Eugène Cortambert
1805–1881

By permission of the Bibliothèque Nationale, Paris

NUMA BROC

1. EDUCATION, LIFE AND WORK

Pierre François Eugène Cortambert was born in 1805 at Toulouse, where his father practised as a doctor. Nothing is known of his youth except that, after his secondary schooling in his native town, he went to Paris while still quite young to complete his education. He was already living there when he was twenty, working as a teacher in private institutions and collaborating in the production of the *Dictionnaire géographique universel* which was published by Picquet. He completed that work alone in 1833.

Deeply interested in geography and education, the primary work of his life was in teaching and popularizing the subject. In 1834 he published his first manual. Soon after this he established a private school for girls in which he was to teach for over forty years and where he was succeeded by his son Richard. He was a friendly, kindly person and a very good teacher, excelling in practical teaching methods and basing much of his work on the use of maps and diagrams. He produced a large number of school manuals, many of which ran into several reprints and new editions. Throughout his life he remained outside academic work in universities; his participation as an assessor in the *Concours Général*, the annual competitive examination in lycées and colleges in Paris and Versailles, was his only contact with state education.

He was a member of the Société de Géographie of Paris from 1836 and became its general secretary for two years (1853–4). In 1854 he joined the geographical section of the Bibliothèque Nationale, and in 1863 on the death of Jomard became its chief librarian.

At the time of the beginning of the Third Republic, Cortambert was one of France's foremost geographers, alongside Vivien de Saint-Martin, Levasseur and Malte-Brun. In 1873 he was elected president of the central committee of the Société de Géographie and so was actively concerned with the organization of the International Geographical Congress held in Paris in 1875. In his later years he collaborated with Ludovic Drapeyron (1839–1901) in the production of the *Revue de Géographie*, but he gradually transferred parts of his work to his son, whose name is sometimes confused with his own and who survived his father by only three years. Eugène Cortambert died in 1881 in Paris, the city he had hardly ever left since he first came to it in his youth.

Cortambert's work in geography lay in several fields. Amongst his contemporaries he was regarded primarily as a teacher and popularizer of the subject. His numerous manuals were adapted to all levels of teaching, from *La petite géographie illustrée du premier âge* (1870) to his *Cours de géographie* (1846) which was adopted for candidates at Saint-Cyr, the military academy. He used his facility for writing to produce many 'specialist' texts, such as *Eléments de géographie* (1828), *Eléments de géographie ancienne* (1834), *Eléments de géographie physique* (1849) and *Eléments de cosmographie* (1851). Compared with contemporary works by Meissas and Michelot and by Abbé Gaultier among others, his manuals were distinguished by the trouble he took to make geography a more interesting subject and to introduce, into the lists of place names and definitions, ideas about botany, geology, ethnography and political economy.

All the manuals were accompanied by small atlases. His *Cours de géographie* was an extremely successful book; it ran through at least nineteen editions between 1846 and 1893 and was translated into Spanish in 1873 for use in the New as well as the Old World. This book was in three parts, of differing length: the first short section covers the general concepts of cosmography, physical geography and ethnography; the second part, which is a little longer, is an overall survey of the five continents; the third part, the major section of the book, consists of geographical accounts, in the form of detailed descriptions, of the countries of the world, largely inspired by Malte-Brun's work.

Among Cortambert's popular works are *Tableau de la géographie universelle* (1832), *Tableau de l'univers* (1848), *Le globe illustré* (1872) and, of special interest, *Les curiosités des trois règnes de la Nature* (1837). In the last the young reader, in the process of discovering the marvels of creation, is led to pay respect to the infinite wisdom of the Creator. This work was reprinted and revised many times.

Pedagogical problems continued to interest Cortambert in his later years. In 1875 at the age of seventy he was elected to preside over the Commission 'Enseignement et diffusion de la géographie' at the International Geographical Congress in Paris. With his colleagues Drapeyron, teacher of geography at the Lycée Charlemagne, and Dupaigne, teacher at Collège Stanislas, he put the case for reform in the teaching of geography: this demanded the co-ordination of history and geography courses, the training of specialist teachers of geography qualified by a special degree, and the development of topography and map reading at all levels of education.

Cortambert also undertook valuable work outside the field of education. As secretary of the Société de Géographie of Paris, he initiated the annual publication of *Notices sur les travaux de la Société*; he was responsible for those of 1852 to 1854. He observed continuously the evolution of his subject; of particular note in this respect was his *Coup d'oeil historique sur les voyages et les progrès de la géographie depuis 1800 jusqu'en 1856*, published in 1856. In the geographical section of the Bibliothèque Nationale, he undertook the preparation of a catalogue of the valuable materials collected by Jomard and thus made this incomparable collection more accessible for study. *Classification de la section géographique* was published in 1877.

Cortambert laid the foundation of a 'geography of intelligence' of France. In 1854 he noted that some regions were more productive of important people than others and that such celebrities seemed to be grouped by regions: the Seine basin had been particularly productive of rulers, the east of soldiers, the south of politicians, Languedoc and Burgundy of intellectuals. *Carte des célébrités de la France* was first published in 1853, and at the International Congress of 1875 he displayed his highly original map which used a range of shade densities to illustrate the extent to which different *départements* had produced famous figures.

Another of his ideas -- to increase the renown of geography in competition with history -- was to endow the subject with a Muse -- *Eugea* -- whom he saw as a young goddess of great beauty yet serious, her head garlanded, looking with an intelligent and profound gaze across space. This vision, which he described in *Parallèle de la géographie et de l'histoire* (1854), embodied his conception of geography,

> ... peignant d'une main habile les paysages et les contrées; assise sur une hauteur lumineuse au voisinage de la mer, d'où elle peut contempler à la fois les deux principaux éléments qui font l'objet de ses descriptions, ayant autour d'elle plusieurs des fruits de ses nobles travaux, des cartes et plans, des livres, un globe, des images des races humaines, quelques uns des instruments qu'elle emploie -- enfin divers produits de l'agriculture, du commerce et de l'industrie.

2. SCIENTIFIC IDEAS AND GEOGRAPHICAL THOUGHT

Cortambert's contribution to geographical thought is to be found not in his manuals, which were generally traditional in concept and form, but in his more original works, in which his contemporaries seem to have taken little interest. When he published *Physiographie* in 1836, Cortambert was aware that he had produced a work that was both difficult and audacious. How could he, a simple geographer, dare to penetrate the scientific secrets of such scholars as Arago, Cuvier, Beaumont or Jussieu? Nevertheless he deemed it desirable to bring together in one volume the isolated general concepts which were to be found separately in specialized treatises on astronomy, geology, botany and meteorology. Ten years before Humboldt's *Kosmos* was published (1845-6), Cortambert was striving to bring geography out of its isolation and to enrich it through contact with the physical, natural and human sciences.

At that time geography was essentially a descriptive subject; he sought to establish it in French schools as a systematic subject. His aims had parallels in other countries. In England, Mary Somerville's book, *On the connexion of the physical sciences*, published in 1834, presaged her *Physical geography* (1848) which broke away from the traditional style of geography manuals. Cortambert's connection with these movements in other countries is suggested by his translation of a text book by J.E. Worcester, the American lexicographer, geographer and historian (1784-1865): *Abrégé de la géographie sacrée* (1830). Alas, Cortambert was no Humboldt and his *Physiographie* was colourless and lacking in originality; he was unsuccessful at that time in reforming other geographical writers who persisted in their listing of places and features as the expression of their subject.

Cortambert's attempts to enliven geography as a school subject and to deepen it as a subject for study were comparable with those of William Hughes, another English writer (1817-76), who was also closely connected with geography in school education and the training of teachers, and a prolific writer of geography manuals and atlases. In 1877-8, Cortambert contributed the preface to Hughes's *Grand atlas universel*, published in Paris by J. Rothschild.

Cortambert thought that geography was always incorrectly assigned in the general classifications of sciences. Some classified it with mathematical sciences, others with historical or social sciences. According to his view, stated in *Place de la géographie dans la classification des connaissances humaines* (1852), geography belonged neither to the physical nor to the human sciences, though it came under both headings. Geographical sciences had their own unity and identity and a distinctive and independent place; along with the economic sciences, they were transitional between physical and moral sciences. Cortambert suggested that there should be a new group of physical-moral sciences, in which geography would stand side by side with ethnography, topography, statistics and economics.

On several occasions Cortambert found it necessary to make a case for the individuality of geography in relation to history. For him history was the knowledge of the past and geography that of the present. Historical geography was but one aspect of the subject and it had given geography as a whole its reputation as a dull subject. In contrast, the 'modern geography' of Cortambert's time was an attractive subject because of its descriptive and imaginative aspects. It might be less philosophical than history, but it was of wider application and a more positive subject for study:

> La terre est la propriété de l'humanité, tout membre de celle-ci est intéressé à connaître ce domaine ... et à savoir quelles productions il peut offrir ... quels sont les lieux qui les fournissent, par quelles voies on peut se les procurer et quels rapports nous unissent avec les autres habitants du globe.

3. INFLUENCE AND SPREAD OF IDEAS

In Cortambert can be discovered the geographical theoretician as well as the teacher and writer of popular geographical books. His epistemological concepts may now appear to be limited but bearing in mind the contempt in which the majority of nineteenth-century thinkers -- such as Ampère and Auguste Comte -- continued to hold geography, one must pay tribute to his attempts to give the subject its rightful place in French schools, colleges and universities. In a century pervaded by the influence of history, Cortambert indefatigably defended modern geography. Although his influence may have been somewhat diffuse, it can be seen as part of the evolution of geography throughout many countries, both through his manuals and his articles, and as part of the full development of scientific geography in France after 1890.

Bibliography and Sources

1. REFERENCES ON E. CORTAMBERT

Levasseur, E. 'Notice nécrologique de E. Cortambert', *Bull. Soc. Géogr. Paris*, vol I (1881), 239-42

Blanc, E. 'E. Cortambert, géographe', *Ann. Soc. Lett. Sci. et Arts des Alpes Marit.*, vol VIII (1882) 312-17

Maunoir, C. *Rapport sur les progrès de la géographie*, Paris, vol II, 1896, 415-16

Polybiblion. Revue bibliographique universelle, 2nd ser, vol XIII, 364 (Paris, Soc. Bibliographique)

Dictionnaire de biographie française, Eugène Cortambert, vol IX, 735, Paris, 1961

Freeman, T.W. *A hundred years of geography*, London, 1961

Broc, N. 'Cortambert et la place de la géographie dans la classification des connaissances humaines (1852)', *Rev. d'Hist. des Sci.*, vol 29 (1976), 337-45

2. SELECTED BIBLIOGRAPHY OF WORKS BY E. CORTAMBERT

1832 *Tableau de la géographie universelle, ou description générale de la terre*, Paris, 526p.

1834 *Eléments de géographie ancienne*, Paris, 228p.

1836 *Physiographie. Description générale de la Nature*, Paris, 336p.

1837 *Les curiosités des trois règnes de la Nature*, Paris (10 eds up to 1884), 318p.

1846 *Cours de géographie, comprenant la description physique et politique et la géographie historique des diverses contrées du globe*, Paris (13 reprints 1852-77), 650p.

1848 *Tableau de l'univers*, Paris, 269p.

1852 *Place de la géographie dans la classification des connaissances humaines*, Paris, 8p.

1853-4 'Notices sur les travaux de la Société de Géographie et des progrès des sciences géographiques', extrait from *Bull. Soc. Géogr. Paris*, Nov., Dec. 1853, 44p., and 1855, 40p.

1854 'Parallèle de la géographie et de l'histoire', extrait from *Bull. Soc. Géogr. Paris*, 1854, 12p.

1854 *Note sur la carte des célébrités de la France*, Paris, 12p.

1856 'Coup d'oeil historique sur les voyages et sur les progrès de la géographie depuis 1800 jusqu'en 1856', extract from *Géographie universelle de Malte-Brun*, 68p.

1856-60, 1863-4 Revision of *Géographie universelle de Malte-Brun*, Paris

1872 *Le globe illustré...*, Paris, 119p.

1876 *Histoire des progrès de la géographie de 1857 a 1874*, Paris, 142p.

1876 'Densité des forces intéllectuelles des diverses parties de la France. Distribution des personnages célèbres', (from *Bull. Soc. Géogr. Paris*, August 1875), 8p.

1877 *Classification de la section géographie*; Bibliothèque Nationale, Paris

Numa Broc is a lecturer in geography at the Centre Universitaire de Perpignan, France. Translated by Marguerita Oughton.

Dates	Life and career	Activities, travel, fieldwork	Publications	Contemporary events and publications
1805	Born at Toulouse			
1821				Founding of the Société de Géographie de Paris
1825	Teacher of geography in various private schools in Paris	Collaborator in the *Dictionnaire Géographique Universel* (publ. C. Picquet and completed in 1833)		
1832			*Tableau de la Géographie Universelle*	
1834	First texbook published		*Eléments de géographie ancienne*	
1836	Member, Société de Géographie, Paris		*Physiographie*	
1837			*Curiosités des trois règnes de la Nature*	
1846			*Cours de Géographie* (first edition)	Publication of Humboldt's *Kosmos*
1852			*Place de la géographie dans la classification des connaissances humaines*	
1853-4		General secretary of the Société de Géographie de Paris		
1854	Joined Geographical section of Bibliothèque Nationale			
1856-60 (also 1863-4)			Revision of *Géographie Universelle* (Malte-Brun)	
1863		Librarian, Geographical Section, Bibliothèque Nationale		
1864				L'Enseignement spécial, created by Victor Duruy (forerunner of 'modern education')
1870				Franco-Prussian war
1871				Congrès de géographie d'Anvers
1872			*Le Globe illustré*	Rapport General sur l'Enseignement de l'Histoire et de la Géographie: E. Levasseur and L.A. Himly

Dates	Life and career	Activities, travel, fieldwork	Publications	Contemporary events and publications
1875		Organizer of Congrès International de Géographie, Paris	*Histoire des progrès de la géographie de 1857 à 1874*	Publication of first volume of Géographie Universelle by Elisée Reclus
1881	Died in Paris			

Charles Andrew Cotton
1885-1970

JANE M. SOONS AND
MAXWELL GAGE

The impressive quantity of written work (more than 160 scientific papers and books) produced by Sir Charles (C.A.) Cotton and the widespread publication of many of these works is in itself an indication of the importance the scientific world has accorded this rather reserved and modest geologist. In those countries where geomorphology is largely the concern of geographers he had an influence on their teaching and thinking almost as great as that of W.M. Davis himself. His textbooks, notably *Landscape as developed by processes of normal erosion* (1941), *Climatic accidents in landscape-making* (1942) and *Volcanoes as landscape forms* (1944), continue in demand more than twenty-five years after their initial publication. Reasons for this are not hard to find in the course of even a cursory review of his writing. Closer study reveals a fascinating development of thought and a precision of expression of that thought which could well be emulated by more modern writers.

1. EDUCATION, LIFE AND WORK
Seeking reasons for a lifelong interest in landforms it is tempting to suggest that Cotton's boyhood travels around the New Zealand coasts and beyond in merchant ships captained by his father may well have sown the first seeds of interest in the diversity of scenery. His university education was, however, in the traditionally 'hard-rock' schools of mining and geology at Otago University in Dunedin, and although a dynamic view of landscape was apparent in some of Cotton's earliest geological writings, Dr. J. Marwick is probably correct when he suggests, in an obituary notice,

that the special problems of teaching geology in the field in the Wellington district seventy years ago would inevitably have led to some emphasis on local landform studies.

In those earlier days there was an almost total lack of any secular basis for geochronology. Cotton however was quick to perceive the abounding evidence in landscapes on both sides of Cook Strait for sequential phases in the evolution of both coastal and inland features. Later he was to recognize the interplay of tectonic and climatic factors which, though at first seeming to confuse and obscure interpretation, actually held the key to a proper timing of geomorphic events. It is not surprising, then, that Cotton at the outset of his career as a geomorphology teacher would have welcomed the promise of order and sequence at that time offered only by W.M. Davis's collations and rationalizations of the pioneering work of late nineteenth-century American geologists.

2. SCIENTIFIC IDEAS AND GEOGRAPHICAL THOUGHT
Wherever his books and papers have been read Cotton has been regarded as one of the foremost exponents of the Davisian approach to the study of landforms. His considerable admiration for W.M. Davis was apparent in comments in a number of papers where either implicitly or explicitly (as in his discussion of the morvan theory in 'Atlantic gulfs, estuaries and cliffs', *Geol. Mag.*, vol 88 (1951) 113-28) he accepts Davis's explanation of a specific piece of landform evolution. In his own mind, however, he considered that he had developed along somewhat different lines. In a

typically firm but gentle reproof he pointed out to one of the present writers '... you have unfortunately failed to note my persistent variation from literal Davisianism (though I hope I have never denied Davis any of the credit due to him...).   I owe an enormous debt to Davis, for all the geomorphology I was able to acquire in my early days came from reading his admirable papers.'   (J.M. Soons, personal communication).

The 'variation from literal Davisianism' lay in Cotton's rejection, after a due trial, of a rigid concept of successive cycles and denudation chronology in the Davisian sense, as applied to the New Zealand landscape.   In spite of this, to the average reader a paper or a textbook by Cotton presented a virtually classic discussion of the evolution of landscape in terms of a sequence of youthful, mature and old forms.   The discussion is usually, if not always, illustrated by clear and apposite block diagrams and by well-chosen photographs of appropriate examples, including many from New Zealand.   This framework formed the basis of several of his papers, emphasis being placed on the hypothetical, deduced evolution of a particular landform or assemblage of landforms. The problem having been outlined -- for example, a question of coastal classification, or of the impact of volcanic events on a pre-existing surface -- its development was meticulously followed through a comprehensive series of possible steps, each stage being so clearly indicated that the reader could not lose the train of thought (though he might be wearied by the effort to follow the numerous ramifications).   In his earlier writings particularly, Cotton was preoccupied with the cyclic concept of landform evolution and with placing in this framework both his deductive sequence and the examples he used to illustrate it. Whatever the experience that initiated the writing of a paper, the discussion of the actual landscape frequently seemed to be of secondary importance to the intellectual exercise of identifying the erosional episode and stage to which it might belong and of placing the example within this cyclical framework.

A careful reading of the first pages of Chapter IV of *Landscape* gives support to the belief that Cotton's 'deviation' was chiefly a swing towards regarding the Davis 'geographic cycle' as an ideal set of models, still useful as a basis for geomorphic analysis and description in terms of a systematic nomenclature notwithstanding the many necessary departures from the oversimplified original.   In other words Davis always, and Cotton initially, may have thought of the famous 'sequential forms' as necessarily having appeared in proper order, but for Cotton from the time when he resumed writing vigorously in the early 1940s each pictured stage and form is offered as a model for convenience in description of landscapes.

This concern with concepts showed a keenly analytical mind as well as one which had considerable imaginative powers in its ability to envisage developments from an initial, given state.   The quality of Cotton's intellect is shown broadly in many of his later papers, which were frequently reviews of works of other writers, drawn together and critically assessed in relation to a particular problem or in an attempt to resolve mutual inconsistencies.   Such bibliographical reviews include discussions of land-

forms of tropical regions ('The theory of savanna planation', *Geography*, vol 46 (1961), 89-101; 'Plains and inselbergs of the humid tropics', *Trans. R. Soc. N.Z. (Geol.)*, vol 1 (1962), 269-77) and a critical review of work on high Pleistocene shorelines ('The question of high Pleistocene shorelines', *Trans. R. Soc. N.Z. (Geol.)*, vol 2 (1963), 51-62).   Such papers were not only the outcome of the very wide reading which characterized the whole of Cotton's career but also models of cool, incisive and unprejudiced comparison of evidence and interpretation.   Moreover they say much for Cotton's ability to envisage the landscape of areas which were personally unknown to him.

Until Kathleen C. Boswell's translation of Walther Penck's *Morphologische Analyse* appeared in 1953 the only readily available summaries in English of the contemporary German schools of geomorphological thought were to be found in Cotton's *Landscape as developed by processes of normal erosion* which came out in 1941. Cotton was unable to accept the empirical rigid tectonic assumptions which underlie the Penckian interpretations of landscape evolution, but his presentation of them was fair and his counter-reasonings easy enough to follow.   They were independent criticisms of the German approach rather than a defence of Davis, whose own reactions to Penck's elaborate, special theories Cotton considered had been less moderately expressed.

In contrast to his attitude to Penck's application of his own principles in interpreting the geomorphic development of western Europe, Cotton was clearly ready enough to promulgate the same basic doctrines of slope-retreat when they appeared in more rational forms with the backing of (among others) his own former student, Lester King.   In the between-war years one of the main points of attack on the simple cycle scheme was the importance given to the ideal case of rapid, uniform uplift of the land, coupled with the assumption that the time covered by each episode of base-level shift was insignificant compared with the length of the succeeding stillstand in which a new cycle was to work out its partial course.   In the strongly tectonic regions of the world neither condition is an appropriate basic premise.   This clearly had its effect on Cotton's thinking, even while he appreciated that in the more extensive stable regions negative eustatic sea-level shifts have been the major cause of base-level changes during the period of evolution of near-coastal landscapes.

Cotton's textbooks were perhaps his greatest contribution to geomorphology.   They had, and still have, a very considerable attraction for a wide range of readers, partly because of their clarity and illustrations but even more because of their convincingly authoritative and logical approach, within the framework of the Davisian cyclic concept.   There are few uncertainties, and the development of landforms and their relationship to the cyclic concept are clearly explained.   The lack of information on process nature and rate, now regarded as one of the major weaknesses of the Davisian approach, does not intrude nor did Cotton apparently regard it as important.   Possibly he did not recognize it as a weakness.   Considerable care is shown in the use of language and in particular in the elaboration of precise definitions.   An absorption with definitions is constantly evinced, and Cotton certainly trained many students in the importance of exact use of terms. It was, perhaps, overdone and did not always have the intended effect -- for example, the suggestion in *Geomorphology: an introduction to the study of landforms* (1942) that a distinction be made between 'cirque' and 'corrie' in the description of glaciated regions, the first to apply to valley-head features and the second to those hanging along valley sides. Reference to *Climatic accidents in landscape making*, which appeared in the same year, suggests that Cotton did not intend such definitions to be as mandatory as successive generations of under-graduates have sometimes thought. The emphasis on definition may at times have obscured the reality of the form described, in much the same way as the meticulous presentation of deductive schemes in his research papers may have obscured a very real desire to understand and explain the perceived landscape.

Cotton never fully accepted as valid the post-1950 developments in geomorphology, with their swing to quantitative measurements and the application of statistical techniques. He did, however, show throughout his career a concern with new developments, both in the body of facts and in concepts, in those areas that appeared to him to be most relevant to the holistic study of landforms. Thus in some of his later papers, and notably in the posthumously published *Bold coasts* (1974), he had no hesitation in referring back to and publicly correcting statements made in earlier works and now considered to be inaccurate or misconceived. The developments in radiometric tech-niques for absolute dating were a significant source of such re-assessments. In early papers Cotton assumed a very long period of post-glacial time in which pro-cesses operated slowly, and this assumption required considerable mental readjustment when the present 10,000-year span was established.

Even so, he seems not to have assimilated (and probably he had little opportunity to assess) modern work on the nature and rates of various processes. This is demonstrated in a number of papers on coastal landforms, in which he appears understandably puzzled by apparent variations in rates of coastal denudation and even finds it necessary to postulate a sharp and inexplicable cessation of cliff recession in the post-glacial period compared with a 'vigorous coastal development and cliff recession in the penultimate cycle' 'Seacliffs of Banks Peninsula and Wellington', *N.Z. Geogr.*, vol 7 (1951), 103-20). The difficulty, still not completely resolved, of reconciling present-day process rates, as observed over short time-spans, with the tempo of long-term geomorphic evolution in varying tectonic and climatic situations was fully appreciated by Cotton, who remained to the end rather cynical about what he thought were largely delusions of precision and objectivity.

3. INFLUENCE AND SPREAD OF IDEAS

More than three-quarters of Cotton's scientific pub-lications are on geomorphology and closely related topics. The courses in physical geography which he gave for B.A. students at Victoria College, Wellington, long before the establishment of a separate geography department must have been among the earliest efforts to establish formal university instruction in the subject in New Zealand. But Cotton's identification at all

times was predominantly with geology, though he re-ceived honours from geographical institutions on more than one occasion.

The Davisian approach to the study of landform evolution is no longer well regarded and many of the examples selected by Cotton to illustrate his arguments, taken from his familiar New Zealand sur-roundings, have been invalidated by recent work which takes into account modern dating techniques, studies of processes and knowledge of the great complexities of Pleistocene climatic change. Nevertheless, his papers and textbooks provided a grounding in obser-vation and analysis of landforms for many of those workers who have contributed to the obsolescence of his work, and they still stand as models in style and presentation. His critical reviews, written in his later years, are well worthy of study and are in some cases the most easily accessible sources of a wide range of material.

Charles Cotton's rather shy and retiring nature was often seen as a barrier between himself and his students and contemporaries. To some extent this was true, and considered together with the undeniable claim that his lectures were seldom inspiringly delivered it lends support to the view that the great impact that he made on geomorphology in the first half of the twentieth century was achieved through his writings. At the same time, those who were fortunate enough to claim a close acquaintance with him saw Cotton as a kind and friendly person, deeply interested in the work and welfare of his students and scientific associates, and ready always to discuss their problems and to share with them his vast knowledge of geological writings in many languages. In his own country he was for a long period the prime authority on geomorphology and a com-prehensive and authoritative student of its landforms. In the wider field of international geography he was again an authority, a scientist able to bring a trained and imaginative mind to bear on some of the world's finest landform examples, and from them to draw inspi-ration and examples for the application of a general theory of landform evolution.

Bibliography and Sources

1. OBITUARIES AND REFERENCES ON C.A. COTTON
Gage, M. *Memorial to Charles Andrew Cotton, 1885-1970*, Geol. Soc. Am., 1971, 9p.
Marwick, J. 'Charles Andrew Cotton K.B.E., 1885-1970', *Proc. R. Soc. N.Z.*, vol 99 (1971), 100-5

2. WORKS BY C.A. COTTON
A complete bibliography exclusive of reviews, corres-pondence and newspaper articles is given in M. Gage, *Memorial to Charles Andrew Cotton, 1885-1970*.
A composite bibliography including reviews and notes up to 1966 is to be found in the following two refer-ences:
Adkin, G.L. and Collins, B.W. 'A bibliography of New Zealand geology to 1950', *N.Z. Geol. Surv. Bull.*, Wellington, 1967, 244p.

Collins, B.W.   'A bibliography of C.A. Cotton', *N.Z. J. Geol. and Geophys.*, vol 9 (1966), 28-44

Complete list of Cotton's books:

1922 *Geomorphology of New Zealand, Part 1: Systematic: an introduction to the study of landforms*, N.Z. Board of Science and Art Manual no 3, Dominion Museum, Wellington, 466p.;   reprinted by N.Z. Govt. Printer, Wellington, 1926

1941 *Landscape as developed by the processes of normal erosion*, London, 301p.;   2nd ed 1948

1942 *Geomorphology: an introduction to the study of landforms*, Christchurch, 505p. (presented as 3rd ed of *Geomorphology of New Zealand*);   4th ed 1945;   5th ed 1949;   6th ed 1952;   7th ed 1958

---- *Climatic accidents in landscape making*, Christchurch, 354p.;   2nd printing 1947

1944 *Volcanoes as landscape forms*, Christchurch, 416p., 2nd ed 1952

1945 *Earth beneath: an introduction to geology for readers in New Zealand*, Christchurch, 128p.

---- *Living on a planet*, Christchurch, 47p.

1955 *New Zealand geomorphology: reprints of selected papers 1912-1925*, N.Z. Univ. Press, Wellington, 281p.

1974 *Bold coasts*, ed B.W. Collins, Wellington, 354p.

Jane M. Soons is Professor of Geography at the University of Canterbury, Christchurch, New Zealand; Dr Maxwell Gage was formerly Professor of Geology at the University of Canterbury and is now an honorary Research Assistant in Geology at Canterbury Museum, Christchurch, New Zealand.

Dates	Life and career	Activities, travel, fieldwork	Publications	Contemporary events and publications
1885	Born in Dunedin, New Zealand			
1889-1903	Attended Christchurch Boys' High School			
1904	Entered Univ. of Otago School of Mines			
1905				*Geology of New Zealand* (P. Marshall)
1907	Graduated B.Sc., Sir George Grey Scholarship; awarded (but declined) Senior Scholarship in Physics			
1908	M.Sc., first class hons. in geology; director of Coromandel School of Mines (until 1909)			
1909	Lecturer in charge of first dept. of geol. Victoria College (University of New Zealand), Wellington		M.Sc. thesis 'Geology of Signal Hill, Dunedin' *Trans. N.Z. Inst.*, vol 41	
1913		Fellow, Geol. Soc. London; Associate, Otago School of Mines (Geology)		
1915	D.Sc. (Univ. of N.Z.)	Collaborated with Weraroa Experimental Farm, Levin		
1916-19		Editor, N.Z. Institute		
1919				*New Zealand plants and their story* (L. Cockayne)
1921	Professor of Geology, Victoria College, Wellington (later Victoria Univ. of Wellington)	Fellow of N.Z. Institute (after 1933 R. Soc. N.Z.)		
1922			*Geomorphology of New Zealand, Part 1: Systematic: an introduction to the study of landforms*	
1926	Married Josephine Gibbons			
1927		Hector Medal, N.Z. Inst.		

Dates	Life and career	Activities, travel, fieldwork	Publications	Contemporary events and publications
1928		Br. Assoc. Advance. Sci. meeting, Glasgow; travel in Europe		*The vegetation of New Zealand* (L. Cockayne)
1930				*Regional geography of New Zealand* (G. Jobberns); *New Zealand birds* (W.R.B. Oliver)
1941			*Landscape as developed by processes of normal erosion*	
1942			*Climatic accidents in landscape making; Geomorphology: an introduction to the study of landforms*	
1944			*Volcanoes as landscape forms*	*Soil erosion in New Zealand: a geographic reconnaissance* (K.B. Cumberland)
1945			*Living on a planet*	
1947		Hutton Medal, R. Soc. N.Z.		
1948		Correspondent, Geol. Soc. Am., later Hon. Fellow		
1951		Victoria Medal, R. Geogr. Soc., London		
1953	Retired from Chair of Geology, Victoria Univ. College			
1954	Hon. LL.D., Univ. of N.Z.; Emeritus Prof., Victoria Univ. College	Dumont Medal, Geol. Soc. Belgium		
1955		Hon. Member, Geol. Soc. Belgium		
1958		Corresponding Fellow, Edinburgh Geol. Soc.		*Climate of New Zealand* (B.J. Garnier)
1959		Knight Commander of the Order of the British Empire		
1966				80th birthday Memorial Issue, *N.Z. J. Geol. and Geophysics* (vol 9, July)
1970	Died at Wellington			
1974			*Bold coasts* (ed B.W. Collins)	

George Davidson
1825–1911

GARY S. DUNBAR

Reproduced through the courtesy of the Bancroft Library, University of California, Berkeley

George Davidson, who has been called 'the greatest scientist to live and work on the Pacific Coast in the nineteenth century', had an extraordinarily long and productive career. After a half-century with the United States Coast Survey, he was appointed Professor of Geography in the University of California in 1898. His voluminous works are of enduring importance to the historical geography and cartography of the Pacific Coast of North America and to the history of Pacific exploration.

1. EDUCATION, LIFE AND WORK

George Davidson was born in Nottingham, England, of Scottish parents in 1825. The family moved to Philadelphia, Pennsylvania in 1832, and George graduated from Central High School -- an institution which was really giving university-level work, despite its name -- at the age of twenty. In 1848 he was awarded a Master of Arts degree by Central High School. After his graduation from the School in 1845 Davidson joined his old patron, Alexander Dallas Bache, the former principal of Central High School who became Director of the United States Coast Survey in 1843. Davidson was employed in the Survey for a full half-century until forced to retire in 1895. Apart from the hiatus of the Civil War, when he was stationed in the East, Davidson spent most of his career on the Pacific Coast, ranging from a base in San Francisco to places as distant as Alaska and Panama. In 1874 he headed the American expedition to Japan to observe the Transit of Venus, after which he continued travelling around the world making observations on irrigation, harbour

improvements, and reclamation projects. He also participated in surveys of irrigation in California's Central Valley.

George Davidson actively supported the scientific and academic institutions of the San Francisco Bay area. He was especially concerned with the affairs of the California Academy of Sciences, of which he was President from 1871 to 1887. When the Geographical Society of the Pacific was founded in 1881, Davidson was called upon to be its President and this he remained to his death thirty years later. The Society, which was said to have been 'the second of its class in the United States' behind the American Geographical Society of New York, was dealt a near-mortal blow in 1906, when it lost its library in the fire following the great earthquake. After the fire the Society's membership declined, and its publications consisted largely of writings by Davidson himself. Davidson built up his own astronomical observatory in San Francisco and was instrumental in the establishment of the famous Lick Observatory near San José. He was also prominent in the affairs of the Sierra Club and the California Historical Society.

The University of California was founded in Oakland in 1868 and moved to its present Berkeley campus in 1873. Davidson was appointed Honorary (*i.e.*, unpaid) Professor of Geodesy and Astronomy in 1870, and he apparently delivered lectures on an occasional basis during the ensuing years. In 1898 he was named Professor of Geography when the Department of Geography was established in the University's new College of Commerce. Although he was then seventy-three years old, eight years beyond the normal retirement age, he travelled to Berkeley three times a week from his home in San Francisco to teach a two-semester undergraduate course, Geography 1-2, 'The currents and climatology of the Pacific Ocean', a course whose title masked its concern with wider topics.   In fact it seems to have been a general course in the physical, economic, and political geography of the world, with emphasis on the Pacific borderlands.   Eye trouble forced Davidson's retirement in 1905, but he continued to teach for two more years as an emeritus professor at a reduced salary.

2. *SCIENTIFIC IDEAS AND GEOGRAPHICAL THOUGHT*

Although Davidson had been engaged in geographical work throughout his professional career, he nevertheless made a great effort to find out what academic geography consisted of at the end of the nineteenth century and to organize his department accordingly. He studied the structure of geography at Oxford and Cambridge and corresponded with Ferdinand von Richthofen of the University of Berlin, who had worked in California with Josiah Whitney in the 1860s. Surveying the literature of geography, Davidson found little agreement on procedures and few organizing principles.   He studied H.R. Mill's *Hints to teachers and students* (1897) and concluded that he would have to forge his own programme for Berkeley's new Department of Geography.

> Dr. Mill ... declares there is no complete summary of Geography, and no single system of teaching it.   He does not say what he considers a summary of geography: and these 'Hints to Teachers' really consist of little more than the titles of certain works -- works which it would require years to read, and more years to digest ... Who is going to make a 'Summary of Geography' from this mass of information and then of what practical use would it be?   ... The labor of gathering together the material to illustrate a single question in Geography is very great.

On the principle that geographical study should begin with areas within reach, Davidson proceeded to initiate his course on the Pacific Basin and to instigate systematic collection of books and maps.   He also urged the creation of a card index of statistical data from government documents.   He even designed map cases himself as there were no satisfactory models to imitate.   To correct geography's deficiency in scientific rigour -- a complaint voiced by contemporary scientists -- Davidson urged that the subject should be expanded to include the principles and practice of geodesy.   The extensive course notes and lists of examination questions that Davidson left behind show that he gave his students a great amount of factual information with a modicum of theoretical or methodological underpinning.

3. *INFLUENCE AND SPREAD OF IDEAS*

Because of the brevity of his academic career and because of the small size of the Department he created, Davidson's influence on academic geography in America was quite evanescent.   His chief contribution to academic geography was the founding of the Department

of Geography in the University of California, the oldest separate department of geography in the United States.   A second appointment to the Berkeley department came in 1901 when Lincoln Hutchinson was made Instructor in Commercial Geography, but many of his courses were taught in the departments of history and economics.   More important was the appointment of Ruliff Holway as Assistant Professor of Physical Geography in 1904.   Holway was the link between the retirement of George Davidson and the arrival of the redoubtable Carl Sauer in 1923, when the Department of Geography was placed on a really firm basis.   In March 1904 Davidson prepared a 'shopping list' of desirable appointments to be made in the Department but only the first of these needs was met:

> Want.
> Instructor in Physical Geography.
> Instructor of the Geography, Productions
>   & Commerce of the Pacific Coast States
>   and British Columbia.
> Instructor in Navigation and Geographic
>   Exploration.
> Instructor in the History of Geography.

It is interesting to speculate about how the Department might have developed if Davidson's plans had been adopted.   If he had been twenty years younger and in good health Berkeley might have had the first truly modern Department of Geography in the United States, instead of having to mark time until the appointment of Carl Sauer.   Today George Davidson's works are useful mostly as sources of factual information to historians and geographers who are concerned with the Pacific Coast of North America.   He will be remembered as a true pioneer of science, not only for his fieldwork and publications but also for his efforts in nurturing the young scientific and academic institutions in the San Francisco Bay area.

Bibliography and Sources

1. *OBITUARIES AND REFERENCES ON GEORGE DAVIDSON*

Wagner, Henry R. 'George Davidson, geographer of the northwest coast of America', *Q. California Hist. Soc.*, vol 11 (1932), 299-320

Davenport, Charles B. 'Biographical memoir of George Davidson, 1825-1911', *Natl. Acad. Sci., Biog. Mem.*, vol 18, 9th mem. (1938)

Lewis, Oscar. *George Davidson, pioneer west coast scientist*, Univ. California Press (1954)

Sherwood, Morgan B. 'A pioneer scientist in the far north: George Davidson and the development of Alaska', *Pac. Northwest Q.*, vol 53 (1962), 77-80

King, William F. 'George Davidson: Pacific coast scientist for the U.S. Coast and Geodetic Survey, 1845-1895', Unpublished Ph.D. thesis, Claremont Graduate School (1973)

2. *SELECTIVE BIBLIOGRAPHY OF WORKS BY GEORGE DAVIDSON*

1859 'Directory for the Pacific Coast of the United States', *Rep. Superintendent Coast Surv. ... 1858*, Washington, D.C., 297-458

1864 *Ibid.*, revised (2nd) ed, *Rep. Superintendent Coast Surv. ... 1862*, Washington, 268-430

1869 *Ibid.*, 3rd ed published as *Coast Pilot of California, Oregon, and Washington Territory*, Washington, 262p.

1869 'Report of Assistant George Davidson relative to the resources and the coast features of Alaska Territory', *Rep. Superintendent Coast Surv. ... 1867*, Washington, 187-329

1869 *Coast pilot of Alaska. (First Part), from Southern Boundary to Cook's Inlet*, Washington, 251p.

1876 'Irrigation and reclamation of land for agricultural purposes, as now practised in India, Egypt, Italy, etc. 1875', 44th Congress, 1st session, *Senate Executive Document* 94, 73p.

1889 *Coast pilot of California, Oregon, and Washington* (4th ed), Washington, 721p.

1902 'The tracks and landfalls of Bering and Chirikof on the northwest coast of America', *Trans. Proc. Geogr. Soc. Pac.*, 2nd ser., vol 1, 1-44

1902 *The Alaska boundary*, San Francisco, 235p.

1904 'The glaciers of Alaska that are shown on Russian charts or mentioned in older narratives', *Trans. Proc. Geogr. Soc. Pac.*, 2nd ser., vol 3, 1-98

1907 'The discovery of San Francisco Bay ...', *Trans. Proc. Geogr. Soc. Pac.*, vol 4, 1-153

1908 'Francis Drake on the northwest coast of America in the year 1579 ...', *Trans. Proc. Geogr. Soc. Pac.*, vol 5, 1-114

1910 'The Origin and meaning of the name California ...', *Trans. Proc. Geogr. Soc. Pac.*, vol 6, part 1, 1-50

G.S. Dunbar is Professor of Geography at the University of California, Los Angeles.

Dates	Life and career	Activities, travel, fieldwork	Publications	Contemporary events
1825	Born in Nottingham, England			
1832	Moved to Philadelphia			
1845	Graduated from Central High School and entered U.S. Coast Survey			
1850	Arrived in California			
1859			*Directory for the Pacific Coast of the United States* ('Coast Pilot', 1st ed)	
1860	Returned to East Coast			
1861				Civil War (until 1865)
1867	Returned to West Coast	Canal survey in Panama, reconnaissance in Alaska		
1868				Founding of University of California
1869			*Coast pilot of Alaska*	
1871	President of California Academy of Sciences (until 1887)			
1874		Transit of Venus expedition to Japan, followed by travel through Asia and Europe		
1878				Coast Survey renamed Coast and Geodetic Survey
1881	President of Geographical Society of the Pacific (until 1911)			
1889			*Coast pilot of California, Oregon, and Washington* (4th ed)	
1892				Sierra Club founded
1895	Retired from Coast Survey	Honorary Vice President of the International Geographical Congress (London)		
1898	Professor of Geography, University of California			Founding of Department of Geography, University of California
1902			'The tracks and landfalls of Bering and Chirikof ...'	

Dates	Life and career	Activities, travel, fieldwork	Publications	Contemporary events
1904			'The glaciers of Alaska ...'	
1906				San Francisco earthquake
1907	Stopped teaching		'The Discovery of San Francisco Bay'	
1908		Awarded Daly Medal of American Geographical Society	'Francis Drake on the Northwest Coast of America'	
1910			'The origin and meaning of the name California'	
1911	Died in San Francisco			

Eratosthenes
c.275 - c.195 B.C.

GERMAINE AUJAC

Eratosthenes was the first author to give the name 'geography' to one of his writings and also the first to find a scientific way of measuring the circumference of the earth. His calculation of the extent of the globe was accepted as accurate for centuries and his map of the inhabited world, which for the first time made use of a network of co-ordinates, inspired many map makers of a later time. A famous figure of the ancient world, Eratosthenes is less well known now for none of his own work has survived except through the summaries, appreciations and critiques of contemporary and later writers. Fortunately these make it possible to assess the contribution of Eratosthenes to geographical thought.

1. EDUCATION, LIFE AND WORK

a. The versatile scholar

Eratosthenes, the son of Aglaos, was born at Cyrene during the 126th Olympiad (between 276 and 273 B.C.) if the view expressed in the *Suidae Lexicon* is accepted but several years earlier according to Strabo (64 or 63 B.C. - 25 or 26 A.D.) who said that at Athens Eratosthenes studied under Zeno of Citium, the founder of Stoicism (*c*. 335 - *c*. 263). Eratosthenes apparently came from a family in comfortable circumstances and had an excellent education. Cyrene, in his day a prosperous town, was the main meeting place for people from the interior of Africa and the Mediterranean basin. When the Ptolemy family came to power in Egypt, the citizens of Cyrene accepted their suzerainty and Magas, brother-in-law of the reigning Ptolemy Soter,

became viceroy in 308. Later, in 275, he tried to become a ruler independent of Egypt and assumed the title of King; the engagement of his daughter Berenice with Ptolemy Euergetes in 260 with the marriage in 246 reunited the two countries.

At Cyrene Eratosthenes learned much of Homer from Lysanias the grammarian, at once an erudite scholar and a devoted admirer of the poet. There is a theory that Eratosthenes was also a pupil of Callimachus (*c*. 310-*c*. 240) who, like Eratosthenes, was a native of Cyrene but settled in Alexandria. If this theory is true, Eratosthenes received a mathematical training from some of Euclid's disciples. But he was still a youth when he first went to Athens. He was immediately impressed by the intense intellectual activity he found there and by the rich diversity of the philosophical doctrines. The Stoician school founded by Zeno of Citium about 300 attracted him at one time but he soon turned to the successors of Plato and attended the New Academy founded by Arcesilaus and directed by him from 268 to 241. Arcesilaus, originally from Pitane in Eolide, had himself been the first pupil of the famous mathematician Autolycus, who was the author of *The sphere in motion* and *The rising and setting of the stars*, the earliest scientific textbooks still preserved. Through Arcesilaus Eratosthenes acquired the best astronomical and geometric teaching of the day.

Strabo noted that Eratosthenes was interested in many differing views and people, among whom was Ariston of Chios, a dissenting Stoic who had been a pupil at the famous school of Arcesilaus, known for his contempt of physical science and for his lectures of compelling charm and also Bion of Boristhenia, the lover of an

exotic style with so many figures of speech that he was
called the 'philosopher in an embroidered cloak'.   In
Athens as in Alexandria later on the diversity of
interests possessed by Eratosthenes astonished his
contemporaries.

b. *The director of the library*

Eratosthenes had already attracted notice in Athens
when Ptolemy Euergetes entrusted him with the education
of the son born to Berenice the Cyrenian in 244.   This
enterprise was doomed to fail as prince Philopator was
a weak-willed individual who, when king from 221-204,
left the government to ambitious but incapable favour-
ites stigmatized by Eratosthenes in *Arsinoe*, written
after the death of the unfortunate sister-wife of
Philopator.   In the other task given to him, to be
Head of the library at Alexandria, Eratosthenes was remarkably successful: he held this post from 230 to his
death at over eighty years of age, probably in 195.
There he was able to draw on a rich bibliographical
mine, to live among the finest scholars of the day in
literature and science.   His varied genius was revealed in poetry, grammar, philosophy, mathematics and
chronology and his knowledge was so universal that he
became known as the second Plato.   His major works all
date from the time spent in Alexandria.   These include
The measurement of the Earth, in which his method was
explained and a *Geography* in three volumes with a map
of the inhabited world.   His other works include
mathematical studies dealing with the duplication of
the cube, a list of constellations in the *Catasterisms*,
reflections on the calendar and the exact length of the
solar year in the *Octaeteris* with a critique of a work
on the same theme by Eudoxus of Cnidus, several writings on grammar, rhetoric and literary criticism and
finally a philosophical work, *On good and evil*.

c. *Personality*

From the range of his works, Eratosthenes would seem to
be the antithesis of a specialist, though in fact he
was a specialist in many fields.   Capable of rigorous
logic, he loved discussion and especially paradox; a
well-informed philologist, he was always dubious about
the subtleties of literary commentators and laughed at
those who looked for scientific teaching in poetry;
interested in philosophical doctrines, he committed
himself to none of them.   Three centuries later Strabo
saw in these characteristics an 'inconsistency of
judgement' and with some acerbity he stressed the contrast between the dedication of the scholar and the
dilettantism of the philosopher or the man of letters;
'perhaps he really wishes to provide himself with some
recreation from his other studies, as a distraction or
just simply a game' (*Geography*, Book I, chapter 2,
paragraph 2).

Possibly it was this capacity for distraction, for
detachment, that enabled Eratosthenes to view his own
researches as from a distance and so to achieve fine
results as a geographer.   Though geography was undoubtedly a science with a dependence on rigorous
logic, it could only advance -- in the state of knowledge and of techniques of Eratosthenes' time -- by
bold approximations which were generally the result of
judicious intuition.   On this Strabo wrote that:
   Frequently Eratosthenes digresses into discussions too scientific for the subject he is

dealing with, but the declarations he makes
after his digressions are not rigorously
accurate but only vague, since, so to speak,
he is a mathematician among geographers and
yet a geographer among mathematicians; and
consequently on both sides he offers his
opponents occasions for contradiction.
(II, I, 41).

Even so, logic and intuition, clarity of synthesis, an
alert and constantly critical spirit, independence of
judgement, all seem to mark Eratosthenes as a man of
genius.   He died at an advanced age from voluntary
starvation, induced by despair at his blindness.

2. *SCIENTIFIC IDEAS AND GEOGRAPHICAL THOUGHT*

A. *The Scientific Contribution*

Eratosthenes' scientific contribution lay primarily in
mathematical geography, in which his originality is
beyond doubt and secondly in physical geography, in
which he generalized results already acquired and co-
ordinated hypotheses already formulated.

a. *The measurement of the earth*

Observers had long since noticed that at Syene - Assuan
at midday of the summer solstice a well was lit up by
the sun to its bottom and that an obelisk cast no
shadow.   Working on the hypothesis generally held at
this time that the terrestrial globe was at the centre
of the universe and could be regarded as a point in the
celestial sphere (so that the rays of the sun are
parallel wherever they fall on the earth), and that
Syene and Alexandria were on the same meridian (in fact
Alexandria is $2\frac{1}{2}°$ west of Syene), Eratosthenes came to
the conclusion that the angle formed with the vertical
by the position of the midday sun in Alexandria at the
summer solstice gave the same angle as the arc of the
terrestrial meridian between Syene and Alexandria.
Gnomonic calculation fixed this angle at one-fiftieth
of the circumference of the earth (7°12', though in
fact 7°7').   The distance between Syene and Alexandria
was estimated -- roughly -- at 5,000 stadia and therefore the circumference of the earth was 250,000 stadia,
which Eratosthenes rounded up to 252,000 stadia to be
divisible by sixty, as any circle was traditionally
divided into sixty segments until Hipparchus introduced
the Babylonian division into 360 degrees.
   This geometric assessment of the earth's circumference had considerable consequences.   By regarding
the stadium, like the metre, as a certain fraction of
the earth's circumference, it was possible to calculate
any arc of the terrestrial meridian, including those
which surveyors were quite unable to measure.   For
example, the obliquity of the ecliptic circle had been
roughly estimated to be equivalent to one side of the
pentedecagon (24°); it was then possible to fix at
16,800 stadia or 4/60 of the terrestrial meridian, the
distance between the equator and the tropics, which
nobody had covered or been able to measure.   Similarly
Eratosthenes located the polar circle at 4/60 of the
meridian (16,800 stadia) from the pole and therefore at
46,200 stadia from the equator, a distance nobody could
test except on artificial terrestrial globes.   It was
now possible to express the latitude of a place in

terms of distance from the equator, and so differences of latitude, obtained through gnomonic methods could be expressed in stadia.   Strabo, for example, observed that Eratosthenes found the distance between Rhodes and Alexandria to be 3,750 stadia, which meant that by using the gnomon he had found a difference in latitude of 5 1/3° (in fact 5¼°) between them.   It was also possible, knowing the length of the equator, to calculate the length of the various parallels of latitude. According to Strabo (I, 4, 6) Eratosthenes estimated that the parallel through Rhodes, at 6/60 from the equator (36°), was 200,000 stadia long.

b. *Map compilation*

Though this was a major contribution to cartography, it was less original than the measurement of the earth and attracted less attention as well as some criticism. Several decades earlier Dicearchus produced an improved map of the inhabited world based on a central axis at approximately 36°N passing through the Straits of Gibraltar, Messina and Rhodes.   This line Eratosthenes used as a reference parallel.   He added as a reference meridian the one based on Syene, Alexandria, Rhodes and the Hellespont.   At Rhodes the parallel and the meridian crossed so this became the central point of the map.   Eratosthenes now went on to map parallels to the north and south of Rhodes and meridians to the east and west.   Henceforth a map was no longer an empirical sketch, or a mere guide book for sailors but an accurate diagram, showing the position of any place in terms of latitude and longitude.   Naturally Eratosthenes was obliged to use some observations of latitude and longitude of doubtful authenticity; this was noticed by later critics who made no useful additions to his work.

In short the map of Eratosthenes marked an advance in cartography, not least because its compiler had given much thought to the problems of map projection from a sphere to a plane surface.   He examined various projections and decided that he would not use one with meridians converging at the Pole because though this was mathematically desirable, the resultant maps were hard to read.   He chose instead an orthogonal projection, simpler to draw, easier to read and involving little distortion in narrow stretches of latitudes. As yet the extent of the inhabited world appeared to be only a comparatively small part of the entire globe: it extended from a parallel halfway between the equator and the northern tropic (nothing was known of areas to the south) to the Polar circle where the explorer and mathematician from Marseille, Pytheas, had located the isle of Thule: in longitude it covered one third of the parallel 36°N so that, as Strabo (quoting Eratosthenes) said (I, 4, 6), 'but for the barrier of the vast Atlantic ocean, it would be possible to sail west from Iberia to India by following the Rhodes parallel for two thirds of the whole circle, as this parallel on which the length of the inhabited world from Iberia to India has been measured is in fact less than 200,000 stadia in total length'.   In all, the known world covered much less than one quarter of the globe and imaginative people, like Crates of Mallus (director of the library of Pergamum) a century later, thought that there might conceivably be three other worlds, like the one already known, in the three other quarters of the globe which presumably existed but as yet were unexplored.

c. *Tidal currents*

Eratosthenes is also famous for his accurate analysis of currents in straits, which he showed to be tidal. Currents in defiles, especially at Chalcis, had always exercised the imagination of observers who failed to find a satisfactory explanation. (Aristotle is said to have thrown himself into the Euripe near Chalcis out of despair at his failure to explain the reverse current.)   Eratosthenes was apparently the first observer who, using the description of oceanic tides made by Pytheas at Cadiz, suggested that there was some connection between oceanic tides and currents such as those through various straits.   He explained that the currents were due to a periodic depression in the water level.   Strabo described Eratosthenes' theory in this passage:

> And Eratosthenes says that this is the reason why the narrow straits have strong currents, and in particular the strait of Sicily which, he declares, behaves in a manner similar to the flow and ebb of the Ocean; for the current changes twice within the course of every day and night and, like the ocean, it floods twice a day and falls twice a day.   Corresponding to the flood-tide, he continues, is the current that runs down from the Tyrrhenian sea to the Sicilian sea as though from a higher water-level -- and indeed this is called the descending current -- and this current corresponds to the flood tide in that it begins and ends at the same time as they do: that is, it begins at the time of the rising and the setting of the moon and it ends when the moon attains either meridian, namely the meridian above the earth or that below the earth; on the other hand, corresponding to the ebb-tide is the return current -- and this is called the ascending current -- which begins when the moon attains either meridian, just as the ebbs do, and stops when the moon attains the points of her rising and setting (I, 3, 11).

Such a pertinent analysis of tidal currents is apparently unique in antiquity.   However Strabo quoted it satirically for, despite the principle of Archimedes, Eratosthenes had dared to suggest a gradient in the level of the sea.

B. *Geographical Ideas*

a. *Geography as a science*

Eratosthenes believed that geography was a science and should be treated as such.   This may now seem to be a banal comment, but it was not the general view at a time when Homer was generally regarded as the chief -- even the only -- geographer and when Apollonius Rhodius, former librarian of the Alexandria library, wrote a long poem supposedly of a geographical character, the *Argonautica*, narrating the Argo's voyage to and from Colchis and the Black Sea. Eratosthenes made constant attacks on writers who interpreted and commented on the writings of Homer and found much in his work that was not there, for he thought that scientific geography could only be written by specialized scholars.

b. *The need for a basis of astronomy and geometry*
Geography rests on an astronomical basis and uses geometrical methods.   Places could only be accurately located by using the astronomer's diopter for the position of stars and of the celestial Pole and also by using the gnomon, whose shadow made it possible to calculate the height of the sun above the horizon and, consequently, to fix the latitude of a place and also the armillary sphere, to the axis of which one could give any position and so imagine the celestial phenomena for any latitude on the earth, even if nobody had ever visited such places as the equator, the terrestrial poles or the southern hemisphere. Geometry is also vital to geography, especially the geometry of the sphere, through which Eratosthenes was able to calculate the size of the terrestrial globe, the length of its great circles, its various parallels and its diameter;   he realized that the earth's relief was of little significance in relation to the globe as a whole.   Representation of parts of the earth on a plane surface also raised geometrical problems which Eratosthenes had to resolve practically by his map making.

c. *Geography and chorography*
Eratosthenes drew a clear distinction between geography as the study of the whole earth of which only a small and favoured part was accessible and known and of the rest about which knowledge could only be speculation, and chorography, a description of the various regions of the known world with their differing characteristics.   Geography as a general science attempted to discern laws applicable to the whole world while chorography on the other hand accumulated facts and, dealing with data and observations, was a form of empirical knowledge rather than a true science.

C. *The World View*

Such a conception of geography as this implied a view of the world not unique to Eratosthenes.   Except for a small number of Atomists and Epicureans, savants and philosophers generally agreed that the terrestrial globe occupied the centre of the celestial sphere and remained perfectly still while the celestial sphere revolved round its axis (and round the earth) from east to west.   The fixed stars appeared to move along parallel circles whereas the planets, the sun, and the moon move along certain oblique circles whose positions lie in the Zodiac.

If the sphericity of the earth was a fact that could be proved by the bending of the horizon or sea level, by the shadow of the earth on the moon during eclipses, and by the attraction of the heavy bodies towards the centre, the spherical shape of the sky was only a useful hypothesis generally accepted by astronomers and philosophers.   Significantly it permeates all former works on the geometry of the sphere and every work on mathematical geography in general: it is a basic hypothesis of the geometrical study of *Phenomena* by Euclid and the two works of Theodosius of Bithynia, *Spherica* and *On geographical places*.

a. *Geocentrism*
All the improvements made by Eratosthenes to mathematical geography are relevant to the geocentric hypothesis and are conceivable only with the view of the world.   But Eratosthenes could not ignore other hypotheses, including those of Philolaus who thought that the earth revolved around some central fire other than the sun, or especially of Aristarchus of Samos who placed the sun at the centre of the universe. Archimedes, who knew the hypothesis of Artistarchus and described it in *Arenaria* still continued to make planetary systems using the geocentric hypothesis. To this hypothesis Eratosthenes remained faithful and it was in fact used by all those astronomers of antiquity who were also geographers.

b. *The earth and the heavens*
Mathematical geography rests, whether it is understood or not, on the assumption that there is a perfect relationship between the earth and the heavens.   The celestial circles project themselves to the earth, determining limits between special categories of phenomena.   The celestial equator is the parallel along which the sun moves on equinoxial days and the terrestrial equator is the circle along which a uniform twelve-hour day can be observed.   If the celestial tropics are the parallels along which the sun moves on solsticial days, the terrestrial tropics divide the areas of the world in which the shadow projected by the sun may fall in opposite directions, either north or south, according to the season, from those areas of the world in which the shadows fall in the same direction all the year round.   The polar circles limit the area around the pole in which the solsticial day lasts for twenty-four hours or more.   In short, the terrestrial globe is a replica of the celestial sphere and for this very reason, Eratosthenes said, the geographer must also be an astronomer.

c. *General laws*
All the researches of Eratosthenes come from the implicit assumption that certain general laws control the functioning of the world but that these laws, simple as they are, may be difficult to discern beneath the clear diversity of the real world.   Eratosthenes therefore ignored insignificant details and used credible approximations, such as 252,000 stadia (divisible by 60) for the world's circumference and 4/60 of a circle for the distance between the equator and the tropics.   He also thought that the various countries should be approximately represented by some form of geometrical figure, for example India by a rhomb and Ariana by a parallelogram.   He constantly sought some explanation for apparently unrelated phenomena, such as currents in straits, oceanic tides or phases of the moon, for he was convinced that any phenomena, such as the flooding of the Nile, must be due to some as yet unknown law.

3. *INFLUENCE AND SPREAD OF IDEAS*
Eratosthenes acquired considerable fame during his lifetime and remained well-known for several centuries. At the famous Alexandrian library, where he worked for almost forty years, he came in touch with the intelligentsia of his time.   There he met famous astronomers,

including Conon of Samos who discovered a new constellation, Coma Berenices, and Dositheos, known for his calendar, as well as geometers of the Euclidian school, and various historians and philosophers. His position as librarian gave him an assured status and the opportunity of co-operation with many scientists of the time, including Archimedes, who wrote to him frequently and dedicated his work on *Method* to him. His own works on geography were widely circulated for more than three centuries and he was regarded as the founder of ancient geography.

Admiration, however, was not invariably given even by critics who owed much to his work. A major critic was Hipparchus, a renowned astronomer, who wrote a three-volume work *Against Eratosthenes*. He argued that Eratosthenes relied too much on approximation and that his map of the known world had few precisely defined points with too much interpolation; he accused him of hasty generalization ignoring crucial evidence of diversity, as in the study of currents in straits. Nevertheless he used the measurements of the earth of Eratosthenes to fix a terrestrial degree at 700 stadia and to design a complete theoretical table of parallels of latitude degree by degree, from the equator to the pole, including the celestial phenomena of each latitude, the duration of the longest day, the height of the sun at the equinox and the solstices and the names of stars permanently visible.

Polybius the historian questioned some of the distances Eratosthenes gave in the eastern Mediterranean, though as Strabo noted he did not provide any better alternatives, but used the idea, put forward by Eratosthenes in the *Geography*, that equatorial regions might be more temperate than tropical regions where the sun was at the zenith for more than forty days round the solstice.

Poseidonius the philosopher thought that a more exact measurement of the earth could be given than that of Eratosthenes by using an astronomical method. His idea was to use the height of a particular star, Canopus, above the horizon in two different places and so to calculate the arc of meridian between these two places; unfortunately he did not make allowance for the astronomical refraction and consequently he gave a false and exaggerated result. Evaluating the difference of latitude between Rhodes and Alexandria at $7\frac{1}{2}^{\circ}$ ($5\frac{1}{4}^{\circ}$ in fact), the circumference of the earth was said to be 180,000 stadia.

Fortunately the successors of Poseidonius, Strabo, Geminos and Cleomedes adopted the measurements of Eratosthenes. Strabo, who was not gifted with a flair for mathematics, chose to develop a regional geography of each country with a human and economic emphasis. He was an eager follower of Eratosthenes, but hardly his academic heir. This Ptolemy became for he cared little for chorography and much for mathematical geography. By his adoption of the estimate of Poseidonius for the earth's circumference, however, he exaggerated the extent of the inhabited world.

Though all of Eratosthenes' work has been lost, some of it has been handed down to posterity in the geographical treatises of Greek and Latin authors. Among the fruits of his work were the conviction of Christopher Columbus that he could reach India by crossing the Atlantic ocean, the concept of measuring length as a fraction of the terrestrial circumference, the use of the orthogonic projection for world maps and the practice of assessing the earth's distributions in terrestrial zones.

By his accumulation of all relevant data and by his organization of these data to provide a general view of the inhabited world, related to the larger terrestrial globe of which it was a part, and by his application for the first time in history of scientific methods of map drawing, Eratosthenes was not only the discoverer of the word geography but also -- and far more significantly -- the founder of geography as a science.

Bibliography and Sources

1. REFERENCES ON ERATOSTHENES
Aujac, Germaine, *La Géographie dans le Monde Antique*, Paris 1975, 128p.
---- 'Erastothène, premier éditeur de textes scientifiques', *Pallas*, vol 24 (1977) 3-24
Paulys Real-Encyclopädie, Knaack, Georg, 'Eratosthenes', vol 6 (1907) col 358-88
Kubitschek, J.W., 'Erdmessung', supp 6 (1935), col 31-54; and 'Karten', vol 10 (1919), col 2022-2149
Lhereux, A. 'La Géographie d'Eratosthène', thèse, Université de Louvain, 1938
Reymond, A., *Histoire des Sciences Exactes et Naturelles dans l'Antiquité Greco-Romaine*, Paris, 1924, 1955 (2 ed)
Solmsen, F., 'Eratosthenes as Platonist and poet', *Trans. Proc. Am. Philol. Assoc.*, vol 73, (1942), 192-213
Strabo, *Geography*, ed H.L. Jones, 8 vols, London and Cambridge, 1949-54
---- *Géographie*, ed G. Aujac, F. Lassere and R. Baladié, Paris, 1966-
Suidas, *Suidae Lexicon Graece et Latine*, ed G. Bernhardy, Paris, 1843
Thalamas, A., *La Géographie d'Eratosthène*, Versailles, 1921, 128p.
Thomson, J.O., *History of ancient geography*, Cambridge, 1948, 427p.
Tozer, H.F., *A history of ancient geography*, Cambridge, 1935, (2 ed), 388p.
Warmington, E., *Greek geography*, London, 1934, 270p.

2. TEXTS AND FRAGMENTS OF TEXTS
Siedel, G., *Eratosthenis Geographicorum Fragmenta*, Gottingen, 1789
Bernhardy, G., *Eratosthenica*, Berlin, 1822
Hiller, E., *Eratosthenis carminum reliquiae*, Leipzig, 1872
Berger, H., *Die Geographischen Fragmente des Eratosthenes*, Leipzig, 1880
Maas, P., *Analecta Eratosthenica*, Berlin, 1883
Bentham, R.M., 'The Fragments of Eratosthenes', unpublished thesis, University of London, 1948

Germaine Aujac is Professor of Greek Language and Literature at the University of Toulouse-le-Mirail. Translated by T.W. Freeman.

Arthur Geddes
1895-1968

ANDREW T. A. LEARMONTH

The only surviving son of Sir Patrick Geddes (q.v.), Arthur was his close supporter to his death in 1932. In the same year he was married at Montpellier to Jeannie Collin with whom he made a home that epitomized his enthusiasm for Scottish-French culture. This, with his devotion to work and a wide range of cultural activities, made him a fascinating personality, a geographer more conscious than most of the wider relations with the natural and social sciences and an enthusiastic supporter of the individuality and personality of Scotland. His main problem as a writer -- and to some extent as a teacher -- was the tangential quality of his mind for in his genuine search for truth he was led into endless related, if perhaps marginal, speculations. He was, however, fortunate in having the guidance of Alan G. Ogilvie, Professor of Geography from 1932-54, as a tactful senior colleague able to assist him to develop a theme in his writing rather than to diffuse his meaning into a maze of possible relationships. At his time geographers were seeking relationships and for Geddes one simple statement was never complete in itself, for it might suggest six related questions, each of which he would be eager to explore. Had he been subject to a more normal school and university education, he might have been better able to focus his attention on one theme but in learning it may be better to strive for eventual synthesis than to achieve analysis of the minute. In choosing, like his father, the former course, Arthur Geddes was a child of his time.

1. EDUCATION, LIFE AND WORK

Few men can have had so unusual, nomadic and unsettled a childhood as Arthur Geddes, and so unconventional an education for the whole family life was dominated by the extraordinary, eccentric, brilliant, erratic and above all demanding father. Alasdair, eight years older than Arthur, was regarded as the brilliant disciple but he might have broken away from his father had he lived. Shortly after Alasdair was killed on war service in 1917, Patrick Geddes's wife died so Arthur was even more closely involved with enterprises in various countries, and particularly in India at the very time when his contemporaries were students in universities. Patrick attracted numerous disciples, including some of great ability, of whom Lewis Mumford was perhaps the most remarkable.

Such a background would have tested any man, and perhaps stimulated revolt. That was not in the nature of Arthur Geddes, who was temperamentally gentle, sensitive and generous, at times delicate in health though physically resilient with a keen enjoyment of mountain walking and -- in his thirties -- running. He was accident-prone as a rock climber and more suited to the mixture of athletic and aesthetic pleasure that mountain walking can provide, though in the countryside he was always appreciative of place, work and folk, the triad of Le Play on which much of Patrick Geddes's thought was based.

At his best Arthur was a brilliant and inspiring teacher, particularly with a small tutorial or seminar group, and with moderate sized classes. Conversation, especially at social dinners, was to him a joy, as it was to the company around him. As a teacher he was at his best after the Second World War. He was often brilliant, memorable, pointedly provocative, though at times unpredictable, erratic and barely relevant. His foibles alienated a proportion of each class but attracted others. In his later years, aware of his detractors, he would on occasion claim as 'his' many able students who went on to achieve eminence in their careers.

His mind appeared to many people to be mercurial, for he was a supporter of so many good causes, including the welfare of Scotland, the maintenance of Gaelic customs and the clan tartans, the translation of Gaelic songs and prayers, the relief of famine in India, the study of its demographic and medical geography (on which some of his best work was done), and the study of the Indian spiritual heritage and outlook, especially as seen in the work of Rabindranath Tagore. His interest in human geography, conceived before the Second World War to be a broad subject susceptible of varied treatment, led naturally to an increasingly practical study of town and country planning, but in this he also looked back to the inspiration of his father whose vision of a master plan for the world was epitomized in the Outlook Tower, the museum of regional planning on the Royal Mile in Edinburgh.

2. SCIENTIFIC IDEAS AND GEOGRAPHICAL THOUGHT

Arthur Geddes was concerned with the unity of all knowledge and lived at the time when many prominent geographers were exploring human aspects of the subject in breadth, looking for generalizations based on studies in depth. Among them were Elisée Reclus, Jean Brunhes, Vidal de la Blache, Maximilian Sorre in France and Carl Sauer in the United States. The regional geographers of the day were working on large areas, such as Reclus on India and Jules Sion on China and surrounding lands. It was India rather than other parts of Monsoon Asia that attracted Geddes most and in his course on the regional geography of Asia he thankfully enlisted a younger colleague to lecture on China. In India he was a friend of many geographers including George Kuriyan and Shiba Prasad Chatterjee. His work on Indian population was written with a fine appreciation of the physical background, acquired by study and through fieldwork. He saw the variability of Indian rainfall as a hazard that could bring disaster to many communities through drought or flood, and was a keen student of disease, including malaria, as a check on population increase. From French geographers he acquired a sensitive understanding of migration as a cathartic human experience. The love of India developed during his residence there from 1921-4, and when his father settled in Montpellier and began his Scots College, Arthur, under the guidance of Jules Sion, then Professor of Geography at Montpellier, wrote his Docteur ès Lettres thesis, *Au pays de Tagore*. This was published by Armand Colin in 1927 and was described by O.H.K. Spate as 'probably the best geographical (and historical) analysis of any area in India'.

From 1938 to 1939, Geddes visited India again and this further experience, fortified by his work for the Ph.D. of Edinburgh University (awarded in 1935) led to a series of papers in various journals from 1938 to 47. Among these were studies of the variability of population increase, on which he later applied his Indian experience to the United States when he was Visiting Professor at Berkeley, University of California in

1952. This work was published in the *Geographical Review* in 1954. In this paper he acknowledged the welcome help of expert statisticians. In his main work on India, population variability was a recurrent theme. He was deeply conscious of hazards such as droughts and floods on Indian rural life and in 1960 contributed a study of the entire Indo-Gangetic lowland to the Institute of British Geographers. The study of population led inevitably to medical geography and he was a corresponding member of the International Geographical Union's Commission on Medical Geography in the years 1949-56. Some of his inspiration came from the work of C.A. Bentley in 1916 on the inverse relationship between population growth and the then prevalent endemic malaria and he was also deeply interested in the work of Cyril Strickland, the malariologist who linked the disease with deltaic landforms. Much of his work was marked by a pragmatic compassion for others. It also reflected his agnostic humanism with a Christian and Hindu background. He was gentle, pacifist, but implacable and immovable on occasion and generally left wing in sympathy, for instance during the Spanish Civil War, though he was repelled by the arbitrary brutality and bureaucracy of Stalin's U.S.S.R. He saw geography as potentially linked with other social and field sciences, and applicable to benefit mankind.

Essentially a world citizen, Geddes was also an ardent, even nationalistic, Scot. His published work on the geography of Scotland included several interesting articles on its small towns and villages, some of which combine an historical approach with a clear appreciation of the qualities of site and the style of architecture. During the Second World War he joined the staff of the Department of Health for Scotland and fortunately was able to develop his interest in regional planning, some of it reflected in papers on rural Scotland and in his book on Lewis and Harris published in 1954. With L.D. Stamp in 1953 he published a 1:625,000 Vegetation Reconnaissance Survey of Scotland, based on work done by various surveyors from 1900 onwards, including Marcel Hardy, the brothers R. and W.G. Smith, with officials of the Forestry Commission and workers of the Land Use Survey, and incorporating his own survey of Lewis done with Marcel Hardy in 1936. The map was somewhat critically received but it was and remains useful as a basis for further work, not much in evidence so far. To an extent unusual among geographers, Geddes contributed articles to non-academic journals, including several dealing with traditional Gaelic songs.

3. INFLUENCE AND SPREAD OF IDEAS

Although Geddes always made an impact on others as a 'character', the shadow of Sir Patrick seemed to some to be over-prevalent at times, especially when so many of his speeches at conferences included some reference to the Outlook Tower. But even those who disagreed -- and on occasion that might mean almost everybody -- were stimulated by his vision of the possibilities of survey, in the field and in the library, as a prelude to replanning. All this was derived from the fundamental *Place, work, folk (Lieu, travail, famille)* of Frédéric Le Play, so basic a concept to Sir Patrick. Numerous students enjoyed the aspiration so clear in

Arthur Geddes's mind, however struggling it might appear in expression. Theories of determinism, and crude phrases like 'geographical control' were antithetical to Geddes and he was therefore much in sympathy with the French geographers of his day, including de Martonne, Demangeon, Sorre, Dresch and later with younger Frenchmen such as Flatrès. With all these he had interesting contacts but he responded particularly to the resurgence of medical geography, favoured by Sorre and especially by Jacques May, a French medical man who became a geographer through his earlier work on Indo-China. Although Arthur Geddes's work on man and disease-cycle relationships in India may be criticized in detail, his great theme of the variation of population trends over time, his insight and grasp of the grand generalization remains fresh and vital to this day. Probably his finest contribution to geography was his work on India though he also wrote much of interest on Scotland and encouraged many others to do so.

At all times he had many devoted friends but his endless outpouring of ideas was unwelcome to some people, especially those of pragmatic mind. Geddes himself inclined to the view that precise definition could be inadequate for it left no marginal area in which the enquirer was led to consider relationships with social and other human problems. He was, for example, critical of the precise analysis of conurbations given by C.B. Fawcett as compared with the broad view, suggestive rather than conclusive, given by Sir Patrick in his book of 1915, *Cities in evolution*. But in that, as in so much else, he revealed his enduring loyalty to his father's singularly wide-sweeping mind, though this loyalty was accompanied by realism about Patrick's damaging impact on his own personality.

Bibliography and Sources

1. OBITUARIES AND REFERENCES ON ARTHUR GEDDES
Watson, J. Wreford, 'Arthur Geddes, D ès L, Ph.D.'
   Scott. Geogr. Mag., vol 84 (1968), 127-8
Learmonth, A.T.A. and Learmonth, A.M. 'Arthur Geddes
   -- an appreciation', *Bombay Geogr. Mag.*, vol 19
   (1970-1), 1-3
'The life and work of Professor Sir Patrick Geddes,
   Part II -- His Indian reports and their influ-
   ences', *J. Town Plann. Inst.*, vol 26 (1940),
   191-4
Introduction to Mairet, P.A., *Pioneer of sociology.
   The life and letters of Patrick Geddes*. London,
   1957, xvii-xx

*2. SELECTIVE AND THEMATIC BIBLIOGRAPHY OF WORKS
BY ARTHUR GEDDES*

a. Research publications
1927 *Au pays de Tagore; la civilisation rurale du
   Bengale occidentale et ses facteurs géographiques*,
   Preface de A. Demangeon. Paris and Montpellier,
   235p. Also issued as *La civilisation ....
   géographiques*. Thèses, Montpellier: Imprimerie

de la manufacture de la charité, 1927. Doctoral
   thesis, Collège des Ecossais, Montpellier.
1928 'Village life in the Eastern Pyrenees' (notes on a
   record of enquiries made by the Social Studies
   Group of the Le Play House Tours Association party
   in the High Pyrenees, August 1927), edited by
   Arthur Geddes, Group Leader, *Sociol. Rev.*, vol 20,
   89-104
1929 'The regions of Bengal', *Geography*, vol 15, 186-98
1935 'Human Geography of Bengal', Thesis, Edinburgh
   University (typescript)
1936 'Lewis', *Scott. Geogr. Mag.*, vol 52, 217-31,
   300-13
1937 'Spain: some geographical factors in the crisis',
   Scott Geogr. Mag., vol 53, 110-20
1937 'The population of Bengal, its distribution and
   changes; a contribution to geographical method',
   Geogr. J., vol 89, 344-68
1938 'The changing landscape of the Lothians, 1600-
   1800, as revealed by old Estate Plans', *Scott.
   Geogr. Mag.*, vol 54, 129-43
1938 'India: (i) The Chota Nagpur Plateau and its
   bordering plains; (ii) The delta of Orissa,
   population and agriculture', *C.R. Congr. Intern.
   Géogr.*, Amsterdam, vol 2, *Trav. Sec.* III c,
   365-96
1939 'Dynamic problems of geographical reconnaissance
   in India', *J. Madras Geogr. Assoc.*, vol 14, 49-58
1941 'Half a century of population trends in India: a
   regional study of net change and variability,
   1889-1931', *Geogr. J.*, vol 98, 228-53
1942 'The population of India: variability of change
   as a regional demographic index', *Geogr. Rev.*,
   vol 32, 562-73
1943 'Variability in rates of population change with
   reference to India, 1881 to 1931 and 1941: some
   statistical considerations', *Indian J. Medic.
   Res.*, vol 21, 115-23
1945 'The foundation of Grantown-on-Spey, 1765', *Scott.
   Geogr. Mag.*, vol 61, 19-22
1945 'Burghs of Laich and Brae', *Scott. Geogr. Mag.*,
   vol 61, 38-45
1947 'The development of Stornoway', *Scott. Geogr.
   Mag.*, vol 63, 57-63
1947 'Rural communities of Fermtoun and Baile in the
   Lowlands and Highlands of Aberdeenshire, 1696:
   a sample analysis of "The pollable returns",'
   Aberdeen Univ. Rev., vol 32, 98-104
1947 'The social and psychological significance of
   variability in population change; with examples
   from India, 1871-1941', *Hum. Relations*, vol 1,
   181-205
1948 'Geography, sociology and psychology: a plea for
   co-ordination, with an example from India', *Geogr.
   Rev.*, vol 38, 590-7
1948 'Conjoint-tenants and tacksmen in the Isle of
   Lewis, 1715-26', *Econ. Hist. Rev.*, 2nd ser.,
   vol 1, 54-60
1951 'The homestead and hamlet in Celtic lands: their
   significance in agriculture, pastoralism and
   fishing', *C.R. Congr. Intern. Géogr.*, Lisbon,
   1949, vol 3, *Trav. Sec.* 4, 453-71
1951 'Changes in rural life and landscape, 1500-1950',
   British Association, *Scientific Survey of Southeastern Scotland* ed. C.J. Robertson, Edinburgh,
   126-34

1952 Report to the Commission of Medical Geography, in
First Rep., IGU Comm. Medical Geogr., Washington
D.C., 17-27

1953 'Variability in population change in India and
regional variations therein 1921-40', *Indian
Geogr. J.*, vol 28, 69-73

1954 'Variability in change of population in the
United States and Canada 1900-51', *Geogr. Rev.*,
vol 44, 88-100

1955 *The Isle of Lewis and Harris; a study in British
community*, Edinburgh University Press, 1955
(Edinburgh University Publications: Geography
and Sociology, No. 2), 340p.

1959 'Scotland's "statistical accounts" of parish,
county and nation: circa 1790-1835/45', *Scott.
Stud.*, vol 3, 17-29

1960 'The alluvial morphology of the Indo-Gangetic
plain: its mapping and geographical significance', *Inst. Br. Geogr. Trans. Pap.*, no 28,
253-76

1960 'The landscape pattern in the Scottish Mid-
Lowland; Lothian and Mannan; its landforms,
climate and ecology and its land use in change',
Prospect, R. Incorporation Archit. Scotland Q.,
no 122, 34-40

1962 'The Royal four towns of Lochmaben: a study in
rural stability', *Dumfriesshire and Galloway
Nat. Hist. and Antiq. Soc., Trans. J.*, vol 39,
83-101

b. Works on planning

1938 'Regional planning and geography in the United
States' *Town Plann. Rev.*, vol 18, 41-7

1938 'Sociology and town planning (Problems in East
and West), *J. Town Plann. Inst.*, vol 24, 358-64

1943 (with A.G. Ogilvie) 'Rural Scotland after the
war', in Gutkind, E.A. (ed) *Creative demobilisa-
tion after the War*, London, vol 2, 32-46

1944 *National parks, a Scottish survey*, Report by the
Scottish National Parks Survey Committee (Chairman: Sir J. Douglas Ramsey) Oct. 3, 1944 (Cmd.
6631) (Geddes was the Survey Officer)

1949 (with F.D.N. Spaven) 'The Highlands and Islands',
in Daysh, G.H.J. *et al., Studies in Regional
Planning*, London 1949, 1-54

1949 'The agricultural unit: the farm-labour team',
Plann. Outlook, vol 1, 5-21

1952 'Resources and prospects of the Isle of Lewis'
and Harris' agriculture, fishing; Harris Tweed;
country arts; team; community and society.
Twelve articles intended to form the concluding
chapters of *Isle of Lewis and Harris* (1955) but
not included in the published work. Reprinted
from the *Stornoway Gazette*, 11 July, *ff*, for the
Outlook Tower Association, Castlehill, Edinburgh
1952

1953-8 *A reconnaissance map of the vegetation of
Scotland 1:625,000* Ordnance Survey (1953/54).
Explanatory text to the planning map, 1:625,000
no 8. *Vegetation: reconnaissance survey of
Scotland (1958)*, Chessington, 1-16

c. Other publications

1951-61 *The songs of Craig and Ben*, vol 1, Edinburgh
1951, 50p.; vol 2, Glasgow, 1961, 90p.

1951 'Scots Gaelic tales of herding deer or reindeer.
Traditions of the habitat and transhumance of
semi-domesticated 'deer' and of race rivalry',
Folk Lore, vol 62, 296-311

1951 'Weather through the seasons on Crag and Peak, as
sung in Gaelic songs', *Weather*, vol 6, 243-6

1961 *Presenting Tagore in sound and sight*, 12 songs
with Tagore's music, translated from the Bengali
and edited.... by A. Geddes, with an article by
A. Geddes, 'Rabindranath Tagore -- bard musician
and seer', in *The Scotsman, Weekend Magazine*,
Exhibition, Edinburgh Festival (1961), 29p.

*Andrew Learmonth is Professor of Geography at the Open
University.*

Dates	Life and career	Activities, travel, fieldwork	Publications	Contemporary events
1895	Born, Ramsey Gardens Edinburgh, 31 Oct., 3rd child of Patrick Geddes and Anna Morton (elder children Alasdair, 4; Norah, 8, later married to Sir Frank Mears, architect and planner)			
1899-1919	Patrick Geddes held Chair of Botany at Dundee, summers only, much travel in winter, children also nomadic, several Scottish homes, winters with aunt in Edinburgh	Geddes children educated at home on principle and greatly influenced by their mother, Anna. Arthur spent some months in Paris as a child and this gave him a lifelong interest in France		Patrick Geddes much in demand as biologist but increasingly as town planner, Ireland, India, Cyprus, etc. University of Jerusalem and elsewhere in Palestine
1910-12	Short spells in schools with tutors, school leaving certificate with 'crammer' rather late because of nomadic existence and rather poor health	A spell of convalescence and open air work at the farm of Mr. J.A. Campbell, Barbreck House, Lochgilphead, Argyllshire, gave him an insight on rural life. His twelve months with Alec Miller, a sculptor of Chipping Campden, who remained a lifelong friend and whose teaching gave Arthur unusual visual skill in conceiving (and drawing) block diagrams as sculptured landscapes		
1913-14	About a term of study Aberdeen University, interrupted by war and father's appeal for help. Pacifist, against most family convictions	Served with the Society of Friends Reconstruction Unit in northern France, providing temporary housing		(1914-18) First World War
1917	Death in action of Alasdair Geddes			
1921	First visit to India of crucial importance to subsequent work when drawn to geography	Patrick Geddes appealed to Arthur to come to India to assist in the Sociology Department, University of Bombay. He became attached to Tagore's circle at Shantiniketan, Bengal, and helped in his father's planning work		Influenza pandemic 1919-21, severe in Deccan towns and villages
1921-4		Work with Tagore's rural construction projects at Srinikatan (near Shantiniketan) and at Bolpur. This led to his later work on 'population stagnation'. A colleague in this Indian enterprise was L.K. Elmhirst, later well known as the founder of the Dartington Hall project and school		

Dates	*Life and career*	*Activities, travel, fieldwork*	*Publications*	*Contemporary events*
1925	To Montpellier to assist his father, with Collège des Ecossais project	Jules Sion (author of *Asie des Moussons* in the Geographie Universelle series) then Professor of Geography at Montpellier University, encouraged Arthur Geddes to write a thesis on the area around Shantiniketan. Despite demands of his father and problems of the college, he gained the D-ès-L with high commendation, and his thesis was published by Armand Colin	*Au pays de Tagore* (1927) (D.ès.L. thesis)	For health reasons Patrick Geddes retired to Montpellier in southern France, and launched grandiose scheme for Collège des Ecossais to marry northern and southern Mediterranean cultures
1927–8	To geography department Edinburgh University (A.G. Ogilvie, Reader, later Professor was head of the department)	Teaching human geography and a regional course on the Far East. Continuing interest in Scotland especially in Lewis and Harris (following up botanical survey by Marçel Hardy)	1929 *The regions of Bengal*	Deepening world economic crisis increased Arthur Geddes's concern for human condition
Early to mid 1930s	Married Jeannie Collin of Montpellier July 1932. Great granddaughter of French regional geographer Elisée Reclus, who while in exile visited Patrick Geddes in Edinburgh, a visit suggested by Kropotkin	Devoted work for Patrick Geddes's Outlook Tower and concern for the Montpellier Scots College. Maintained and developed his interest in Gaelic poetry and culture. Resumed his work on the population of Bengal, the subject of his Ph.D. thesis of 1935. Wrote and spoke on the contentious issues of the Spanish Civil War	'Lewis', *Scott. Geogr. Mag.*, 1936	Patrick Geddes died April 1932
Late 1930s		Return visit to India (1938–9)	'Spain: some geographical factors in crisis', *Scott. Geogr. Mag.*, 1937 'The population of Bengal', *Geogr. J.*, 1937	Spanish Civil War 1936–9
1941–45	Temporary Civil Servant	Wartime service with Department of Health for Scotland. Some fieldwork on possible National Parks for Scotland. Indian work developed further	'The population of India: variability of change as a regional demographic index', *Geogr. Rev.*, 1942	Second World War (1939–45). Approach of Indian Independence
1946	Return to Edinburgh geography department	Member of the I.G.U. Commission on Medical Geography (1949–68), for which he provided reports	with F.D.N. Spaven, *The Highlands & Islands* 1949	Indian Independence and Partition into India and Pakistan 1947
1951	Year at University of California, Berkeley		'Variability in changes of population in the U.S. and Canada' *Geogr. Rev.* 1954 Varied papers on Scotland, India, medical geography	

Dates	Life and career	Activities, travel, fieldwork	Publications	Contemporary events
1952	Return to Edinburgh University, Department of Geography	Worked for a time in the Social Sciences Research Centre, University of Edinburgh. Still eagerly developing the ideas of Sir Patrick Geddes, especially the Outlook Tower: the centenary of his father's birth was celebrated in 1954	Papers on Scotland, especially historical geography	
1955	Visit to India, as visiting Professor with the Indian Statistical Institute. Travelled widely in the sub-continent		*The Isle of Lewis and Harris*, 1955	
1960			'The alluvial morphology of the Indo-Gangetic plain', *Trans. Inst. Br. Geogr.*	
1965	Retired from teaching	In retirement, continued writing and planning further research, to within a few days of his death		
1968	Died 5 April			

Patrick Geddes
1854-1932

W. IAIN STEVENSON

Patrick Geddes was without doubt one of the most
original and productive minds of his time. His work,
like that of Marx, Freud and Darwin, not only brought
about a reorientation and revitalization of academic
enquiry, but also had wide repercussions on popular
thinking. But the influence of Geddes's mind on the
course of geography was more than that of a thinker
working in another discipline. Although not trained
as a geographer, he worked in areas which were or
have become mainstream geography. He was at once a
theoretician and a practitioner, an influence and a
contributor.

1. EDUCATION, LIFE AND WORK
Patrick Geddes, born in Ballater, Aberdeenshire in
1854, was the fourth son of Alexander Geddes, a soldier
in a Highland regiment, and Janet Stivenson Geddes
whose family came from the coalfield town of Airdrie.
Although devoutly Presbyterian the Geddes family was
strict rather than stern, and his childhood was happy.
From the earliest he showed an interest in the natural
environment and after his family moved to a rural
cottage outside Perth (in 1857) his incipient interests
in nature were allowed full play. Encouraged by his
father he learned a great deal about botany and became
a competent field naturalist. In these pursuits he
displayed the energy that was to characterize him
throughout his life. In search of botanical specimens he frequently climbed nearby Kinnoul Hill. From
there, the view over the Tay floodplain is uninterrupt-
ed and impressive and it was an image that was to stay
with Geddes. When he later formulated his concept of

regional survey he emphasized the necessity of beginning with an overview from a vantage point; 'Outlook
Towers' were invariable features of the Museums and
Institutes he designed and he explicitly illustrated
his famous 'Valley Plan of Civilisation' with a section
from Kinnoul Hill to the North Sea.
   In 1874, Geddes enrolled at Edinburgh University
to study botany, but he was appalled by the rigidity
and lifelessness of the course and left within a week.
He was not however completely disillusioned by academic
botany and, after reading *Lay sermons*, he went to study
under T.H. Huxley (1825-95) at the Royal School of
Mines in London. Although Geddes was later to dis-
agree strongly with Huxley's 'Nature red in tooth and
claw' interpretation of Darwinian evolution, his contributions to the formation of the young man's ideas
were fundamental and permanent. Apart from introducing him to Darwin and giving him a sound training in
scientific method, Huxley encouraged him to read widely
in philosophy and general science, particularly the
works of Thomas Carlyle and Charles Lyell. The strong
feeling for historical evolution which permeates all
Geddes's later work stems directly from his association
with Huxley.
   Geddes also read works of which Huxley did not
approve. The Comtean positivism of Richard Congreve,
Huxley's great rival, impressed Geddes and so did the
political economy of Henry George, which Huxley once
described as 'the damnedest nonsense'. However, the
work of Frédéric Le Play (1806-82) the sociologist,
which Geddes first encountered during a visit to the
Sorbonne in 1878, decisively changed his ideas. Le
Play's triad of *Lieu, travail, famille*, Geddes was to

translate to *Place, work, folk*, the three themes that
became the basis of his sociological investigations,
and were later incorporated into his concepts of city
and region.

Through Huxley's good offices, Geddes was
appointed senior demonstrator in practical physiology
at University College London in 1877 but he left two
years later to organize the zoological station at the
University of Aberdeen.   During this period he began
to travel widely, visiting France (1878-9), Italy
(1879) and Mexico (1879).   The visit to Mexico proved
crucial to the course of Geddes's career.   During the
course of his stay, he contracted eye trouble and,
threatened by blindness, was forced to spend several
weeks in a darkened room.   Since he could not pursue
botanical fieldwork, his thoughts turned to Le Play's
sociology and its implications.   His sight restricted,
he developed the folded paper and diagrammatic models
('thinking machines') that he used later to illustrate
his lectures and writings.   When he returned to
Scotland to become lecturer in zoology at the University of Edinburgh in 1880, his enthusiasm for the role
of objective scientist was already waning and he had
begun to develop an abiding interest in the problems
posed by society and environment.

A growing political liberalism also marked his
new attitude.   His family background, particularly on
his mother's side, had disposed him towards a certain
radicalism but after 1880 he turned progressively towards socialism.   Friendships with the anarchist
Pyotr Kropotkin, whom he first met in London in 1876,
and the brothers Reclus, Elie and Elisée, were doubtless important in introducing him the libertarian ideas
of the time.   He read Ruskin, Marx and Morris and,
although unconvinced by the more extreme arguments, remained a lifelong democrat.   He refused a knighthood
in 1912 and only accepted it in 1932 'as an aid to
carry out schemes against the Philistines who have
long hindered me, but whom it now bluffs and silences
amazingly'.   It was typical of Geddes that his socialism should be distinctly practical (significantly he
disliked the Fabians, particularly Shaw and Wells).
In 1887, he and his wife (he had married the previous
year) moved into the slums of James Court in Edinburgh
and began a programme of rehabilitation and education.
The concept of improvement without community disruption
or wholesale demolition which, termed 'conservative
surgery', became a guiding principle of his planning
ideas first gained expression in this small but extremely successful project.

In 1888, Geddes applied for the chair of botany
at Edinburgh University.   However, the nature of his
extra-curricular activities (along with James Court,
these included organizing students into a self-govern-
ing hall of residence, the convening of a Summer School
in 1887 and the reading of papers on economics before
the Royal Society of Edinburgh) was deemed 'conduct un-
becoming a botanist' by the University establishment,
and although he was the best qualified candidate, he
was rejected.   In the following year he was offered
the newly created chair of botany at University College
Dundee, which he held until 1919.   At Dundee Geddes
was required to teach for only three months of the year
and he was thus able to devote more of his time to his
social science interests.   He continued to research
and publish in biological science (Geddes wrote or

collaborated in four major standard texts, *The evolution
of sex* (1889), *Chapters in modern biology* (1893),
Biology (1925) and *Life* (1931)) but the major part of
his output after 1889 was concerned with the theme of
society and environment.

After 1889, Geddes also became increasingly preoccupied with education at both the theoretical and
practical levels.   In 1890 he published a severe but
just critique of Scottish universities which, hide-
bound in their medieval structure, were extremely
rigid and unresponsive to new ideas.   Although he was
associated with universities throughout his working
life and indeed was actively involved in planning new
educational institutions in India and Palestine, he was
always impatient with the formal procedures, academic
conservatism and exclusivism of most institutions.
This impatience was reflected in his exploration of
alternative means of education.   His earliest essay
in this field was his Summer School which first met in
Edinburgh in 1887 and continued annually there until
the First World War.   He conceived the original meeting as an extension course in 'seaside Zoology and
garden Botany', but as his interests expanded, so did
the range of subjects offered.   By the mid-1890s, the
Summer School had become a major international cultural event and the visiting lecturers included Ernst
Haeckel, Edmond Demolins, the brothers Reclus,
Kropotkin and A.J. Herbertson.

Another of Geddes's experiments in progressive
education was the 'Outlook Tower' opened in Edinburgh
in 1892.   Characterized by Charles Zueblin as 'the
World's First Sociological Laboratory' the Outlook
Tower was in fact a graphic celebration of Geddes's
manifold interests.   It included exhibits from fields
as diverse as chemistry and economics, astronomy and
anthropology.   However, the central feature was the
room devoted to 'Edinburgh and its region', comple-
mented by a panoramic view of the city and its environs
from the tower balcony.

In 1897 Geddes visited Cyprus to study the
Armenian refugee problem.   He encountered conditions
of extreme poverty and deprivation and he organized
several practical aid projects.   These included
agricultural settlements round Larnaca, a sericultural
college at Nicosia and the establishment of a large
collective farm, all of which he largely financed
himself.   In an article in the *Contemporary Review* of
June 1897, he reported on his activities in Cyprus and
made a plea for similar agricultural improvements
schemes elsewhere: 'Solve the agricultural question
and you solve the Eastern question.   Give men hope of
better land, of enough food for their families, and
you remove a main cause of bloodshed' (*Contemp. Rev.*,
vol 71 (1897), 290).

Geddes's reputation as a specialist in social problems was steadily growing and in 1903, he was invited
by the Carnegie Trust to prepare a 'civic survey' of
the city of Dunfermline.   (*City Development, 1904*.)
Although its recommendations were not implemented, the
report helped to establish the nascent subject of town
planning as a respectable and coherent discipline.
Geddes prepared a similar survey of Dublin in 1914 and
between 1916 and 1919, in what is sometimes regarded
as the apotheosis of his career, produced a series of
comprehensive civic surveys and planning reports for
Indian cities.

Characteristically, his concern with urban matters was not only practical but also theoretical and educational. Especially after 1905 urban themes became dominant in the curricula of the Edinburgh Summer School and were more strongly emphasized in the Outlook Tower exhibits. In 1911, a Geddes-designed 'Cities and Town Planning Exhibition' was opened in London, and he was the prime mover in the organization of the first world 'Congress of Cities' at Ghent in 1913. Complementary to these activities, he also produced a series of theoretical and popular writings on urbanism, including 'Civic education and city development' (1905), 'Civics as applied sociology' (1905-6), 'City deterioration and the needs of a city survey' (1909) and culminating in his *chef d'oeuvre, Cities in evolution* (1915).

Although essentially a pacifist, Geddes (like Kropotkin) believed that the First World War presented society with an opportunity to cleanse itself and that a new social order must thereby emerge. To encourage this aim, he launched in association with the social theorist Victor Branford, a series of hortatory tracts entitled 'The making of the future'. Of these, *Ideas at war* (1917) and (in collaboration with G. Slater and V. Branford) *The coming polity* (1917,1919) give the clearest indication of the Geddesian vision of the post-war world, a Utopia of social responsibility and social justice. His bitter disappointment over the terms of the Armistice and his failing health, which had begun to weaken after the death of his wife and elder son in 1917, contributed to his retirement from Dundee in 1919. He did not however seek seclusion. Hoping to rescue some vestiges of his wartime Utopian schemes, he allied himself with the Palestine Zionist movement and was consulted on the plan for Tel Aviv. He also returned briefly to India to organize the School of Civics and Sociology at the University of Bombay.

In 1924 he resurrected an old ambition and, with the help of Paul Reclus (the son of Elie) and the biogeographer Charles Flahault, he established the 'Scots College' at Montpellier as an international study centre. There, surrounded by enthusiastic students (one of whom he married in 1928), he continued to expound and expand his ideas. Some of his last work is among his best, particularly the important paper *'Rural and urban thought'* (1929). The characteristic energy never left him and he was in the midst of plans for an American lecture tour when he died in 1932.

2. SCIENTIFIC IDEAS AND GEOGRAPHICAL THOUGHT

a. The geographical contribution

Geddes's concern for society and environment gives most of his work geographical relevance. Although he variously termed his studies 'sociology', 'civics' or 'geotechnics', the common element of environment unites them all with geography. However, in his inter-relationship of Place, Work and Folk, derived from Le Play, and his insistence on the importance of historical evolution, derived from Huxley, he arrived at a broad integrative viewpoint that many contemporary geographers, still struggling with the rival claims of 'environmentalism' and 'possiblism', found intimidating and difficult to grasp. Geddes in his turn found geographical small-mindedness and bickering tedious in the extreme and although he held many geographers in high esteem and was indebted to their work, it is perhaps not surprising that he rarely wrote on 'geography' or published in geographical journals.

Nevertheless, his contributions to geographical thought were impressive and abiding, perhaps most so in his theoretical and practical consideration of regionalism. From the first, Geddes had stressed that the region was the only context in which the workings of society could be adequately studied. Geddesian regions were not defined arbitrarily, but were real entities, concrete in the experience of the inhabitants. Regional consciousness, he argued, was the major arbiter of regional boundaries. Moreover, regions were interdependent and thus hierarchical. Typically the libertarian, Geddes saw in an hierarchical regional structure the opportunity for regional devolution and the celebration of regional diversity. In 1904, he wrote:

> Regionalism ... begins by recognising that while centralisation to great capitals was inevitable, and in some measure permanent this is no longer completely necessary... The increasing complexity of human affairs ... has enabled the great centres to increase and retain their control; yet their continued advance is also rendering decentralisation, with government of all kinds, increasingly possible... We see, then, that the small city is thus in some measure escaping from the exclusive intellectual domination of grander ones, and is tending to redevelop, not indeed, independence, but culture individuality (*City Development*, 1906, 216.)

A general concern for regionalism is apparent in most of Geddes's writings after 1895, but two specific themes are particularly recurrent. The first is the development and explication of the heuristic device for studying regions, the 'Valley Section' or 'Valley Plan of Civilisation', and the second, the importance and methods of regional survey.

Superficially the Valley Section was a simple empirical generalization. 'The general slope from mountain to sea which we find everywhere in the world ... this way the world is built' (*Survey*, vol 54 (1925), 288). However, the river basin was conceived as more than a neat topographic regularity; it was the focus of society's ('Folk' in Geddesian terms) interaction with economy ('Work') and environment ('Place') and as such was '... the essential unit for the student of cities and civilization'. Furthermore, since '... by descending from source to sea we follow the development of civilization from its simple origins to its complex resultants', the Valley Section illustrated both the evolution of ways of life from upland pastoralism to urban living and their mutual indispensibility. In the Valley Section, Geddes was able to unite Le Play's sociology and evolution in a coherent manner, producing a regional concept which he felt was, in some measure, applicable to geography.

> All things are here. This is no longer our mere school-book, with its image of a country as a coloured space on a flat map, with only 'boundaries' and 'capital' and so on. It is first of all the essential sectional outline of a geographer's 'region', ready for study.

Geographers themselves were not so sure.   Some saw the concept as environmentalist, notwithstanding Geddes's careful stressing of historical factors, while others were wary of the integration of anthropological, socio- logical and economic perspectives.   Even H.J. Fleure, a close personal friend, writing to Geddes on 25 Feb- ruary 1917, said that 'the Valley Section *per se* gives too static an impression' while agreeing with the general principle.

The concept of regional survey was the practical complement to Geddes's theoretical regionalism.   Sur- vey however was not just for specialists; it was an essential pursuit for everyone who was curious about his surroundings: 'Every inhabitant from child to patriarch should strive to know what his region con- tains'.   Curiosity should not be the only motive of regional survey, however.   Geddes hoped that by en- couraging greater regional awareness, survey would also foster a concern for improvement.   'The citizen must first study all these things with the utmost realism, and then seek to preserve the good and abate the evil with the utmost Idealism.'   In regional survey, the egalitarian Geddes thus saw the opportunity for the involvement of as many people as possible in decision making about regional futures.   In a letter to Marcel Hardy on 13 November 1920, Geddes suggested that the skills required were minimal and the concepts easily grasped:

> These general conceptions ... can thus be
> reached, e.g. notably by women, workers,
> artists, etc., without profound scientific
> studies, just as the last generation of
> the people became socialistic without any
> profound economics or history.

Geddes's vision of the region as a living entity is illustrated by his use of medical metaphors to describe the methods of regional survey.   He wrote of survey as 'diagnosis before prescription', of 'conservative sur- gery' and he compared the regional surveyor to 'an old fashioned family doctor' who, although he might make use of specialist reports, would by 'general knowledge of the processes of health and disease reinterpret the specialist view of the problem, and may thus modify the treatment, or change it altogether'.

The main application of regional survey was to regional planning.   In all his planning reports, from the Dunfermline study of 1904 to the Indian reports of 1916-19, Geddes began with a detailed regional in- ventory and analysis, on which basis he prepared his conclusions.   In theoretical works like *Cities in evolution* he continually referred to regional survey as a *sine qua non* for regional planning.   When he established the Regional Planning Association in 1914 it quickly formed a strong alliance with the Town Planning Society, and included many prominent planners, like Patrick Abercrombie and Raymond Unwin.

One of the most stimulating of Geddes's exercises in regional survey, however, was more clearly associ- ated with geography than planning.   In his delightful paper 'Edinburgh and its region: geographical and historical' (1902) he presented a thoughtful and humane essay on the evolution and distinctive charac- teristics of a 'peculiarly complete geographic and occupational environment' (*Scott. Geogr. Mag.*, vol 18 (1902), 302).   As a demonstration of the usefulness of the regional survey method the paper is notably

successful, particularly in its evocation of Edinburgh's 'spirit of place'.   The ideas on which this paper were based had earlier been expressed in abstract form in a paper 'The influence of geographical conditions on social development', in *Geogr. J.* (vol 12 (1898) 580-7) published four years previously: in this paper Geddes speculated on the relationship between societal and regional evolution.   Here, in the study of Edinburgh, he first gave substantive form to his regional survey approach.   It may not be without significance that both these important contributions appeared in geographical journals.

Regionalism and regional survey were the unifying concepts embodied in the design of the Outlook Tower. Part museum and part allegory, it was organized to illustrate and explain the Geddesian view of the world by leading the visitor through the Geddesian regional hierarchy.   The ground floor displays were concerned with the World in general, those on the next, Europe, progressing then to the English-speaking societies, then Scotland and finally Edinburgh.   The culmination of this was the view from the balcony of the tower over Edinburgh and the Forth lowland, which Geddes hoped would stimulate the viewer to relate what he saw to the exhibits he had just seen.   The whole point of the tower was the direct involvement of the spectator, thus making him a practising regional surveyor.   Geddes ex- plained these ideas in the handbook *First visit to the Outlook Tower* (1906) but to many his carefully linked exhibits seemed a random jumble.   However, numerous geographers, including A.G. Ogilvie and R.N. Rudmose Brown, endorsed the methods of the Outlook Tower.   On 9 June 1905 Elisée Reclus wrote to Patrick Geddes:

> The Outlook Tower is as yet the only educa-
> tional museum which addresses itself to the
> (*sic*) man as a whole, which tries to wake in
> him the thoughtful, the emotional and the
> observing instincts alike, and to enhance
> his synergetic value...   May the workers
> fill the Outlook Tower with a real audience.

Reclus then went on to propose a formal association between his Geographical Institute at the *Université Nouvelle* of Brussels and the Outlook Tower.

Geddes himself realized the importance of the methods of the Outlook Tower for geography and in 1902 proposed, with the support of J.G. Bartholomew, the cartographer, Albert Galéron, the architect, and Reclus, a much larger and more sophisticated similar institution as a 'British National Institute of Geography'.   In the *Draft Plan*, Geddes stated his conception of the Institute as a symbolic sythesis of geography: '...it represents the latest of not a few attempts to unite and harmonise many lines of geo- graphical activity and educational endeavour'.   The ground floor was to contain exhibits of 'descriptive and general geography', including two giant globes, celestial and terrestrial, designed by Galéron and Reclus.   However, in typically Geddesian fashion the general exhibits were only to serve as an introduction to the key feature of the building, the 'Tower of Regional Survey':

> But this cosmic presentment of Universal
> Geography, which sets out from the World
> as a whole, involves as its complement,
> the converse or human method.   This is
> presented in the Tower, proceeding from

the visible and immediate prospect. Thence
we descend, storey by storey, through City
and Province and Region or State to Nation
and Empire, and thence again to the larger
Occidental civilisation; and finally to
the Oriental and Primitive sources -- the
facts of geography and history, the problems
and possibilities being also represented, as
far as may be, upon each level (*Scott. Geogr.
Mag.* vol 18 (1902) 143).

Although the project was never realized, it attracted
widespread interest and was publicly supported by prominent geographers and others, including the rather
unlikely figures of Sir Archibald Geikie and Sir
Clements Markham.

Geddes's interest in cities was an outgrowth of
his regional thought. Cities were pivotal in the
Geddesian region, the foci of regional activities,
the cores of regional consciousness. Unfortunately,
Geddes's attitude toward cities has been the subject of
much misinterpretation. He has been accused variously
of attempting to impose rural values on urban areas, or
of overstressing the regional context of cities while
underemphasizing their internal dynamics. To Geddes,
this implied dualism between city and country was meaningless. Urban and rural were not separate, but
rather essential extensions of each other: 'It takes
the whole region to make the city'.

Geddes's major urban manifesto was the long
treatise *Cities in evolution* (1915). For all its dis-
cursiveness, ungainly writing and typically Geddesian
haranguing, it has been judged one of the outstanding
works on urban development and city planning written in
the last two centuries. In the context of its time,
it was also a revolutionary book, especially in its
strong feeling for history and tradition, which directly countered the anti-historicism of contemporary
urbanists, particularly the 'Garden City' school of
Ebenezer Howard. The whole book has relevance for
geographical thought, but the chapter headed 'the
population map and its meaning', a *tour de force* of
theoretical and empirical city-regionalism, is of
particular importance. Here Geddes discusses not
only the geographical form of cities, but also their
dynamics and growth. The chapter bristles with strik-
ing new ideas, among them the now familiar concepts of
'Conurbation' and 'Megalopolis'. Many of the points
raised in *Cities in evolution* were expanded in later
writings, but he never produced a better or more comprehensive statement of his views. It was the most
influential of his works, and bore practical fruit in
the town planning of Patrick Abercrombie in Britain and
Benton Mackaye in the United States, and contributed
to the ideas of the city ecology school of Ernest W.
Burgess and R.E. Park. It is the only one of Geddes's
major works which is still in print.

b. Geographical influences and opinions of geography
Throughout his career, Geddes displayed a certain impatience with academic geography. Benton Mackaye
quotes him as remarking 'Geography is a descriptive
science, it tells what is'. He once remarked 'Geotechnics' (his own approach) 'is applied science -- it
shows what *ought to be*'. His work was nevertheless
directly influenced by geographers. His early reading
had included the works of Carl Ritter and Alfred

Hettner; his biological interests had led him to
Edward Hahn's original theories of domestication, and
he was certainly aware of Alexander von Humboldt's
treatise on biogeography. However, French humanism
particularly as expressed in the geography of Elisée
Reclus appealed more to Geddes than Germanic science.

Reclus (1830-1905) was the author of the *Nouvelle
géographie universelle* (1876-94) -- the first detailed
world geography to be more than a compendium of facts.
Geddes considered him to be the supreme geographer of
his day, to be included among the greatest French
thinkers of the nineteenth century with Louis Pasteur
and Marcelin Berthelot. Geddes recorded his general
debt to the *Géographie universelle* on several occasions
and he acknowledged its specific influence on his conception of the Valley Section. The anarchist philosophy of Reclus, particularly the Bakuninist vision of
world brotherhood and harmony with nature, was attractive to Geddes. Its precepts, which Reclus explained
to Geddes in numerous personal meetings and a pro-
tracted correspondence, shaped Geddes's conceptions of
community and region. Reclus himself was a 'venerated
figure' in the Geddes household and his nephew Paul,
himself a geographer, was adopted as Geddes's protégé.

Another geographer who deeply influenced Geddes
was the Russian Pyotr Kropotkin (1842-1921). Like
Geddes, Kropotkin had been a natural scientist and his
early work in glaciation and orography would be enough
to earn him a permanent place in the history of geography. However, he became dissatisfied with the
social and political detachment of science and, like
Reclus, turned to anarchism ultimately becoming its
most important spokesman and revolutionary theorist.
Geddes and Kropotkin formed a lasting friendship after
the latter fled from Russia to Britain in 1876.
Kropotkin was a regular contributor to the Edinburgh
Summer School and many of his ideas were incorporated
into Geddesian philosophy. Kropotkin's neglected
classic of geographical social criticism, *Fields,
factories and workshops* (1898), inspired many of
Geddes's insights on regionalism, particularly the
concept of regional interconnectedness, while the
arguments of *Mutual aid* (1904), Kropotkin's masterly
refutation of social Darwinism, are clearly discernible
in urban translation in *Cities in evolution*.

Nevertheless Geddes found most of his geographical contemporaries, at least in Britain, shallow and
dull, though he appreciated the breadth of view of
H.J. Fleure, who used his experience of a wide study
of field sciences with a growing interest in archaeology and anthropology as a basis for regional study as
part of geography. Though he wrote little geography,
Geddes taught it and among his papers in the National
Library of Scotland are preserved his lecture notes for
an introductory course of geography given at Dundee in
1898, which, although written in the scrawled diagrammatic shorthand so familiar to students of Geddes's
manuscripts, give a fairly clear idea of the tenor and
orientation of Geddesian geography. The notes open
with a familiar theme: 'See for ourselves! Hill,
tower'. 'Old Geog.?' queries Geddes, 'No, diverse
with three "streams"' (rather) 'a synthetic geography-
evolutionary, local, Scotland and N.W. Europe'
stressing 'the value of microcosm'. But how is such
a synthetic geography to be achieved? 'Develop
senses', Geddes posited, 'emphasis on sight, emotion,

experience'. This was 'real Geog', the 'awareness of locality' utilizing 'even odour, taste and memory'. Thus geography, he contended, was not a catalogue of facts, nor even a science, but a humanity, 'born of appreciation and an ethical sense'.

Although Geddes disliked the practice of geography he had great faith in its potential. Luckily geographical orthodoxy did not stifle geographical thought and eventually a new tradition in British geography began to form in a Geddesian mould. The spread and influence of Geddes's ideas among geographers is traced in the concluding section of this essay but it is opportune here to give some indication of how Geddes saw the relationship between geography and his own work at the close of his career. In February 1932, Fleure wrote to Geddes offering him, at the unanimous invitation of the council, the presidency of the Geographical Association. Geddes delightedly accepted, and in one of his last pronouncements wrote to Fleure on 13 February:

> More and more clearly, all my thinking is on and from our geographical basis, and my graphs are all developed in relation to maps as a shrunken landscape, as a basis for ideas and actions in the *concrete*, and with its parallels and meridians clarifying all presentments of abstract ideas in the sciences.

c. The world view

The Geddesian world view was based on the intertwining of a natural science training with intellectual pantheism (uniting a strong historical sense and a broad mastery of sociology, philosophy, economics and social criticism) and political radicalism. This combination produced his distinctive orientation towards the problems he sought to study. The impact of biological (particularly botanical) ideas is readily discernible in all Geddes's work. He often cast his expositions in organic terms, as in the 'Talk from the Outlook Tower' entitled *Cities and the soils they grow from* (1925). Equally his monism, as expressed for example in his conception of the unity of city and region, owed much to ecological theory. Indeed, Geddes was largely responsible for the expansion of the ecological perspective from biology to social science. In many places, but particularly in the paper *Rural and urban thought* (1929), he proposed the viewpoint that has since become known as 'human ecology'.

Evolution was the other major idea which he translated from biological to social science. However, he abhorred the earlier application of evolutionary theory to society which is incorrectly called social Darwinism (since its arguments have no actual basis in Darwin's theory). Geddes certainly owed much to Huxley, and greatly admired his achievements in biological theory, but he felt that the latter's excursions into social theory, ('society red in tooth and claw' as it were') were perversions. Geddes did not take as narrow an interpretation of Darwinian theory as Huxley for to him Darwin's thought not only implied struggle and competition, but also stressed the existence of co-operation and harmony in the organic world. It was this truly Darwinian perspective that made *Cities in evolution* one of the earliest humanistic applications of the great evolutionist's ideas to social phenomena.

The third legacy from natural science was Geddes's concern for observation and classification before generalization. This Linnaean procedure, first learned on Kinnoul Hill and later, more formally, at the dissecting table was also incorporated into his social thought, notably in regional survey.

His wide reading in the liberal arts and embryonic social sciences together with his political radicalism resulted in a humanism which distinctively tempered his scientific outlook. Particularly influential were the works of Auguste Comte, the ideas of Thomas Carlyle and the writings of the Utopian socialists such as Robert Owen and Charles Fourier. Throughout his life, he sought to eradicate mechanistic and deterministic overtones, although he perhaps never achieved this to his satisfaction; as he wrote to Fleure on 7 June 1915: '(I need) constructive criticism, like yours, with your humanistic standpoint, so complementary to my environmental one'. Despite such protestations, Geddes's standpoint was indeed humanistic, as his sympathy for the individual, his concern for the integrity of communities, his impatience with bureaucracies, and his loathing of metropolization, (for which he coined the telling term 'para sitopolis') all betray. Humanistic ideas are constantly articulated in his writings, perhaps most clearly in the paper, 'Our city of thought' (*Survey*, vol 54 (1925)).

The Geddesian world view welds apparent opposites — scientific objectivity and social commitment, accurate measurement and artistic reflection, conservation of the past and the shaping of the future. To Geddes such a view contained no contradictions. His world was a unity, his world view a synthesis. Man and Nature were a false dualism, as were Art and Science. Only by bringing these ideas together could the world be understood and changed. On this he wrote to Marcel Hardy on 13 November 1920:

> The Survey and Planning of *Place* involves increasingly a survey of Work and of Folk ... (but) Beyond all (this) ... comes the deeper and higher side of Life. That of Emotionalised (and Emotive) 'Dream' and Thought and Imagination — and these towards realisation — in Deed.

3. INFLUENCE AND SPREAD OF IDEAS

Most modern geographers are familiar with the content, if not the authorship, of Geddes's ideas. His views are recognizable in many of the characteristic concerns of what in Britain is termed 'human geography'. Yet, paradoxically, Geddes himself had little to do directly with the formation of the Geddesian tradition in geography. His importance for the discipline lies rather in how he influenced, by personal contact and by his writings, the thinking of geographical contemporaries who then went on to interpret and use his ideas in their own writing and research.

The most enthusiastic protagonist of Geddesian ideas in geography was his younger son, Arthur (1895-1968; q.v.) who taught social and historical geography at Edinburgh University for many years. In published work and teaching, Arthur re-interpreted and illustrated his father's themes of regionalism, drawing heavily on Scottish examples. He was a sensitive writer and a good teacher, but he always remained rather in awe of his father and was thereby perhaps prevented from realizing his full potential as an original thinker. He lacked his father's broad integrative perspective and he tended toward eclecticism, even whimsicality, though he inherited some of the Geddes energy as well as a firm belief in the virtues of field observation. With A.G. Ogilvie he organized numerous geographical field excursions both as adjuncts to teaching and for the general public. Montpellier was a favourite venue. Although Arthur Geddes was limited by an uncritical attitude towards his father and by his own intellectual shortcomings, he helped to introduce a humanistic and field-oriented tendency in Scottish geography, still discernible in the geography curricula of the Scottish universities.

The work of the biogeographer Marcel Hardy (1876-1940) owes much to the tutelage of Geddes senior. The two men probably first met at the Paris Exhibition of 1900 and after Hardy completed his doctorate at the Sorbonne in 1902, he joined the botany staff of University College, Dundee at Geddes's invitation. With a background in Vidalian human geography and botany, he shared Geddes's own breadth of interests and social conscience which became manifested in a concern for the practical applications of Geddesian principles. In 1906 he settled in Mexico, where he organized a 'Colonization Land Project' in the Tezonapa Valley using the 'Valley Section' as a development model. This was among the first successful land colonization and reclamation schemes in Latin America and has been frequently emulated. Similar practical projects dominated the rest of Hardy's career and he became a kind of roving geographical consultant, with such diverse appointments as War Reparations Commissioner in Germany in 1920, agricultural planner in Uruguay during the 1920s, and the chief designer of Lord Lever's model West Highland village at Leverburgh. He maintained close contacts with Geddes, who continually offered encouragement and advice. Although primarily an applied geographer, Hardy also made a substantial contribution to 'pure' geography. His doctoral dissertation, revised with Geddes's help and published in 1905 as *Esquisse de la géographie et de la végétation des Highlands d'Ecosse* remains a classic of biogeography. In collaboration with Arthur Geddes, he produced important studies of Highland settlement, while his textbook, *Geography of plants* (1920), is a pioneer study of plant distribution and ecology.

Another of Geddes's protégés at Dundee was Andrew J. Herbertson (1865-1915) whom he employed as Demonstrator in 1890. Herbertson's earliest work had been in natural science, but his association with Geddes turned his interests more towards social concerns. During the 1890s he lectured on the geography of cities at the Edinburgh Summer School (interestingly enough, in partnership with Elisée Reclus). When he went to Oxford to teach in the School of Geography his relationship with Geddes became less close, but his ideas continued to run on Geddesian lines. His most influential work, on 'natural regions' is in particular clearly derivative of regionalism.

A more direct, but less well-known, geographical application of Geddes's ideas occurs in the work of C.B. Fawcett (1883-1952). Fawcett had studied under Herbertson at Oxford, and taught at Southampton, Leeds and University College, London. He became an enthusiastic advocate of regional survey and Geddes invited

him to speak on the relationship between regionalism and geography at the Dublin Summer Meeting in 1914. Apparently Geddes regarded Fawcett's ideas very highly, for in 1917 he invited him to contribute a book on the geographical aspects of political devolution for 'The Making of the Future' series. This appeared in 1919 under the title *Provinces of England*.

Provinces of England still ranks as one of the major single achievements of twentieth-century British geography. In it, Fawcett proposed a federal structure of twelve autonomous provinces for England, each one closely defined and supported by cogent geographical argument. The book justly made Fawcett's reputation and had great political influence but its ideas are pure Geddes. All six 'principles of the division' were clearly inspired by Geddesian regionalism, particularly so in the organization of each province round a 'definite capital which should be the real focus of regional life', that provincial boundaries should 'be drawn near watersheds, rather than across valleys' and that 'the grouping of areas must pay regard to local patriotism and tradition'. Fawcett may have lacked some of Geddes's *élan* and almost all of his humanity, but in *Provinces of England* he ably translated somewhat vague ideas of regionalism into a workable scheme.

Notwithstanding substantial contributions from Arthur Geddes, Hardy, Fawcett and others like Benton Mackaye and Vaughan Cornish, the major responsibility for the making of the Geddesian tradition in geography lies firmly with H.J. Fleure (1877-1969). For Fleure, 'the sparkling mind of Sir Patrick Geddes' was the inspiration of much of his geographical thought (*Geography*, vol 34 (1949) 7). In his turn, Geddes found in Fleure a mind that equalled his own in breadth of interests and synthetic power, and he always valued Fleure's work most highly. Originally a zoologist, Fleure shared Geddes's background in the natural sciences, but after an introduction to physical anthropology, he too became dissatisfied with the abstract and mechanistic outlook of science. In anthropology and geography, 'the discipline of *actuality*', he saw opportunities for the development of a more satisfying humane viewpoint. Rather like Geddes, his revolutionary ideas were much misunderstood, but their worth was eventually recognized by his appointment both to the chair of Geography and Anthropology at Aberystwyth and to the Secretaryship of the Geographical Association in 1917.

Although traces of Geddes's influence are noticeable in earlier writings, it was with *Human geography in Western Europe* (1918) that Fleure first made a direct and conscious effort to tie geography to Geddes's teaching. Written for the 'Making of the Future' series, the book was intended to complement and amplify with a geo-anthropological perspective the visionary ideas expressed in the accompanying texts, particularly 'Our social inheritance'. It was well received, and he followed it up with other geographical interpretations of Geddesian themes, including a particular influential study of 'Human regions' (*Scott. Geogr. Mag.* vol 35, 1919) and papers on the relationship between geography, regional survey and planning. These writings culminated in 1947 in the publication of 'Some problems of society and environment' (*Inst. Br. Geogr. Trans.* vol 12, 1947) which represented Fleure's mature thoughts on the nature of human geography, and is still perhaps the best summary and justification of the Geddesian viewpoint in geography.

However, the importance of Fleure as the prime interpreter of Geddes in geography and of the vitality of the Geddes-inspired school of human geography he founded, stemmed from his abilities constructively to criticize and expand Geddesian concepts in terms of his own philosophical and intellectual standpoint.   Like Lewis Mumford, Fleure was not merely a popularizer.   Rather he used the ideas as a springboard from which to develop his own original and stimulating perspectives.   The grand systematizations of society and environment, like the Valley Section, seemed too rigid to Fleure (as we have seen, Geddes suspected this, and he sought mitigating humanistic criticism from Fleure) and, although remaining true to their spirit he re-worked them into a more dynamic and less deterministic form, rounding them out with insights from anthropology and archaeology.   In studies like *Natural history of man in Britain* (1951), Fleure comes closer to realizing the Geddesian ideal of humanistic synthesis than Geddes ever did.   Nevertheless the main Geddesian themes clearly provided Fleure with constant inspiration, and despite a differing emphasis and an arguably more sophisticated articulation of ideas he never deviated from the basic postulates of the Geddesian framework.   Thus by expanding, qualifying and developing Geddes's ideas, Fleure rendered them accessible and acceptable to geography.   So much so that today many of them have become incorporated into conventional wisdom or are even obvious commonplaces.   There is perhaps no better testimonial to their greatness.

Bibliography and Sources

The objective of this bibliography is to provide as comprehensive a list as possible of writings by and about Patrick Geddes for those interested in pursuing research on his geographic thought and impact.   This does not mean that everything written about him or indeed by him has been included.   For obvious reasons papers on technical biological topics have been excluded (though the general biological texts are included) as have some other minor ephemera such as his literary journal *Evergreen* and some poetry.   Less defensible, perhaps, is the omission of the original references to the Indian Town Planning Reports;   these reports are of undeniable importance to students of Geddes's ideas, but since they were all prepared locally in very small editions (often typewritten) they are virtually unobtainable today.   Indeed many have apparently disappeared completely and no one even seems at all sure just how surveys were undertaken. Jacqueline Tyrwhitt (1947) has published a judiciously edited selection from most of the extant reports and, as far as can be ascertained, only one report (Lahore) has ever been reissued *in toto* (*Govt. of Pakistan*, 1965;   Stalley, 1972).   For most purposes, Tyrwhitt's still easily obtainable book will suffice until the major research task of locating, editing and publishing the complete run of reports (if this still remains possible) is undertaken.

1. OBITUARIES AND REFERENCES ON PATRICK GEDDES

a. Biographies

The biographies of Defries and Boardman are uncritical eulogies;   Boardman's study in particular verges on the saccharine.   Mairet's study is more balanced and makes good use of primary material but it tends to be rather more of a commentary than an exhaustive biography.   The biography by Abbie Ziffren is based almost entirely on secondary sources and contains numerous errors of fact and inference.   In any serious study of Geddes, all these works must be used with circumspection and care.   Mrs. Kitchen's popular biography achieves a high level of insight.

Boardman, P. *Patrick Geddes, maker of the future*, Chapel Hill, North Carolina, (1944)

Defries, A. *The interpreter Geddes, the man and his gospel*, London, (1927) and New York, (1928), 334p.

Mairet, P.A. *Pioneer of sociology.   The life and letters of Patrick Geddes*, London, (1957), 226p.

Ziffren, A. in M. Stalley (ed) *Patrick Geddes. Spokesman for man and the environment*, New Brunswick, Rutgers University Press, (1972)

Kitchen, P. *A most unsettling person.   A biography of Sir Patrick Geddes*, London, (1975), 351p.

Reilly, J. 'The early social thought of Patrick Geddes', Unpub. Ph.D. thesis, Columbia University, New York, (1971)

Stevenson, W.I. *Patrick Geddes and geography, a biobibliographical study*, Occasional Paper 27, Department of Geography, University College, London, (1975)

b. Obituaries

J.R. Inst. Br. Archit., 4 June 1932, 630: *Mus. J.* vol 32, 1932, 72: *New York Times*, 18 April 1932, 15: *Proc. R. Soc. Edinburgh*, vol 52, (1931-2), 452-4: *Sociol. Rev.*, 'A Sheaf of tributes to the memory of Sir Patrick Geddes' (includes appreciation by Vaughan Cornish) supplement to vol 24, (1932), 351-400: *The Times*, appreciations, 19, 21, 23 and 27 April 1932.

c. Manuscript Sources

The National Library of Scotland, George IV Bridge, Edinburgh EH1 1EW, houses an important collection of manuscript material relating to Geddes.   The bulk of this was presented to the Library in 1961 by Lady Mears (Norah Geddes) and Arthur Geddes.   It has recently been fully catalogued (call nos. MSS 10501-10657) and indexed.   It contains most of Geddes's correspondence from the late 1870s until his death, some personal papers (including notes for lectures and publications), photographs, a portfolio of 'thinking machines' (some of which have apparently never been published and are thus of the utmost interest) and other memorabilia.   A smaller collection of Geddes manuscripts was later acquired by the Library from the estate of Arthur Geddes and is available for study (call nos. Acc. 5796).   This latter series contains some material of great interest to geographers, including letters from R.N. Rudmose Brown and some of Arthur Geddes's early notes and geographical papers.

*d. Critical studies, commentaries and memoirs
relating to Geddes's life and work*

Anon. 'The Valley Plan of Civlisation', *Archit.
Yearb.*, no 12, (1968), 65-71

Anon. 'Actualités de Patrick Geddes textes et documents', *Archit. d'aujourd'hui* no 143, (April
1969), v-vii

Branford, Victor *Citizen soldier: a memoir of Alasdair
Geddes*, London, (1917), 20p.

Barker, Mabel *L'utilisation du milieu géographique
pour l'education*, Librairie Nouvelle,
Montpellier, (1926), (2nd ed Paris, 1931), 249p.

Boardman, Philip *Esquisse de l'oeuvre educatrice de
Patrick Geddes*, Collège Ecossais, Montpellier,
(1936), 203p.

Carter, H. 'The garden of Geddes' *The Forum*, vol 54,
(1915), 455-71, 588-95

Fleure, H.J. 'Patrick Geddes (1854-1932)', *Sociol.
Rev.* (new series), vol 1, (1953), 5-13

Gardiner, A.G. *Pillars of society*, London, Dent (1913),
(pp. 128-34 on Geddes)

Goist, Park Dixon 'Patrick Geddes and the city', *J.
Am. Inst. Plann.*, vol 40, (1974), 31-7

McGegan, Edward 'Sir Patrick Geddes', *Scott. Bookman*,
vol 1, (1935), 99-106

McGegan, Edward (with A. Geddes and F.C. Mears),
'The life and work of Sir Patrick Geddes',
J. Town Plann. Inst., vol 26, (1940), 189-95

Mumford, Lewis 'Patrick Geddes, insurgent', *New Repub.*,
vol 60, (Oct 30, 1929), 295-6

Mumford, Lewis 'Patrick Geddes and his cities in
evolution', *Mag. of Art*, (Washington D.C.),
vol 44, (1951), 25-31

Saisset, P. 'Philosophie et urbanisme -- un précurseur,
Patrick Geddes', *Archit. d'aujourd'hui*, vol 45,
(Sept 1954), n.p.

Tyrwhitt, Jacqueline *Patrick Geddes in India*, London,
(1947), 100p.

Zueblin, Charles 'The world's first sociological laboratory', *Am. J. Sociol.*, vol 4, (1899), 577-92

2. SELECTED WRITINGS OF PATRICK GEDDES

1881 'The classification of statistics and its results',
Proc. R. Soc. Edinburgh, vol 11, 295-322

1884 'Analysis of the principles of economics', *Proc.
R. Soc. Edinburgh*, vol 12, 943-80

1886 'A synthetic outline of the history of biology,
Proc. R. Soc. Edinburgh, vol 13, 904-11
'Theory of growth, reproduction, sex and
heredity', *Proc. R. Soc. Edinburgh*, vol 13,
911-31
*Conditions of progress of the capitalist
and the labourer* Lecture 3, in Oliphant, J. (ed),
'Claims of Labour Series', Edinburgh, 275p.

1887 *Industrial exhibitions and modern progress*,
Edinburgh, 51p.
Every man his own art critic, Manchester, 32p.

1888 *Co-operation versus socialism*, Manchester,
Co-operative Printing Society

1889 (with J. Arthur Thomson) *The evolution of sex*,
London, 322p. (2nd ed, 1901, 342p.)

1890 'Scottish university needs and aims', *Scots
Mag.*, (Aug)

1893 *Chapters in modern biology*, University Extension
Manual, London, 201p.

1897 'Cyprus, actual and possible, a study in the
eastern question', *Contemp. Rev.*, vol 71, 892-908

1898 'The influence of geographical conditions on
social development', *Geogr. J.*, vol 12, 580-7

1900 'John Ruskin as economist', *Int. Mon.*, no 1,
(March), 280-308

1900 'Man and his environment. A study from the Paris
exposition', *Int. Mon.*, no 2, 169-95

1902 'Note on a draft plan for Institute of
Geography, *Scott. Geogr. Mag.*, vol 18,
142-4
'Proposed National Institute of Geography',
Scott. Geogr. Mag., vol 18, 217-20
'Nature study and geographical education',
Scott. Geogr. Mag., vol 18, 525-35
'Edinburgh and its region, geographical
and historical', *Scott. Geogr. Mag.*, vol 18,
302-12

1903 'In an old Scots city', *Contemp. Rev.*, vol 83,
(April), 559-68

1904 (with J. Arthur Thomson) 'A biological approach',
in J.E. Hand (ed), *Ideals of science and faith*,
London, 322p.
*City development. A study of parks, gardens and
culture institutes: Report to the Carnegie
Dunfermline Trust*, Dunfermline, n.p. (1st ed),
Edinburgh, P.G. and Colleagues (2nd ed), 232p.

1905 'The schools at Abbotshowle', *Elementary School
Teacher*, vol 5, 321-33, 396-405
'Civic education and city development', *Contemp.
Rev.*, vol 87, 413-26
'Civics as applied sociology', *Sociol. Pap.*, vol
1, 103-38 (Part 2 in *Sociol. Pap.*, vol 2, (1906),
57-109)
The world without and the world within, London
and Bournville, 38p.

1906 *First visit to the Outlook Tower* Edinburgh, P.G.
and Associates, n.p.
'Suggested plan for a civic museum', *Sociol.
Pap.*, vol. 3, 197-249

1907 'Making of the future', *Sociol. Rev.*, vol 9,
100-4
'Summer in an Old Scots Garden', *Living Age*,
no 254, 620-6

1909 'City deterioration and the needs of a city
survey', *Ann. Am. Acad. Polit. Soc. Sci.*, vol 34,
54-67
'Early homes and haunts of Carlyle', *Oxford and
Cambridge Rev.*, no 8, 11-17

1910 'The civic survey of Ediburgh', *Trans. Town
Plann. Conf. of October 1910*, 537-74 (reprinted
by the Outlook Tower Association, Edinburgh, 1911)

1912 *The masque of learning*, Edinburgh, P.G. and
Colleagues, 90p.

1915 'Wardom and peacedom. Suggestions towards an
interpretation', *Sociol. Rev.*, vol 8, 15-25
Cities in evolution, London, Williams and Norgate.
2nd ed (ed J. Tyrwhitt and A. Geddes) London,
Williams and Norgate, New York, Oxford University
Press 1950, 409p. Another ed, London, Ernest
Benn, 1958, New York, Howard Fertig, 1969. Re-
vised with additions introduction and commentary.
New Brunswick, Rutgers University Press, 1972

1917 (with G. Slater) *Ideas at war*, London 256p.
(with Victor Branford), *The coming polity*, London,
264p., (2nd ed, 1919), 332p.

1919 (with Victor Branford), *Our social inheritance*, London, 318p.

1920 *An Indian pioneer, the life and work of Sir J.C. Bose*, London, 259p.   'Essentials of sociology in relation to economics', *Indian J. Econ.*, vol 3, 1–56 (2nd part, vol 5 (1922), 257–305)

1921 'Palestine in renewal', *Contemp. Rev.*, vol 120, 475–84

1923 'A note on graphic methods, ancient and modern', *Sociol. Rev.*, vol 15, 227–35

1923 *Dramatisations of history*, Edinburgh, P.G. and Colleagues, 90p.

1924 'The mapping of life', *Sociol. Rev.*, vol 16, 193–203

1925 'Talks from my Outlook Tower' series, *Survey*,
  (a) 'A schoolboy's bag and a city's pageant', vol 53, 525–29, 553–54;
  (b) 'Cities and the soils they grow from' vol 54, 40–5;
  (c) 'The valley plan of civilisation', vol 54, 288–90, 322;
  (d) 'The valley in the town', vol 54, 396–400, 415–16;
  (e) 'Our City of Thought', vol 54, 487–90, 504–7;
  (f) 'The education of two boys', vol 54, 571–5, 587–91
  'Huxley as Teacher', *Nat.*, vol 115, 740–3
  (with J. Arthur Thomson), *Biology*, London, 254p.

1926 *Coal crisis and the future*, London, (edited and introduction by P. Geddes), 111p.
  'A national transition', *Sociol. Rev.*, vol 18, 1–16
  'The making of our coal future', *Sociol. Rev.*, vol 18, 178–85. Reissued by Le Play Press, 1926

1927 'The charting of life', *Sociol. Rev.*, vol 19, 40–63

1929 (with Victor Branford) 'Rural and urban thought. A contribtuion to the theory of progress and decay', *Sociol. Rev.*, vol 21, 1–19

1930 'Ways of transition: towards constructive peace', *Sociol. Rev.*, vol 22, 1–31, 136–41

1931 (with J. Arthur Thomson) *Life: Outlines of general biology*, (2 volumes), London, 1515p.

W. Iain Stevenson was on the staff of the Geography Department, University College, London and now works for a firm of publishers.

CHRONOLOGICAL TABLE: PATRICK GEDDES

Dates	Life and career	Activities, travel, fieldwork	Publications	Contemporary events and publications
1854	Born at Ballater, Aberdeenshire, 20 October			
1857	Family moved to Perth			
1859				*On the origin of species* (Charles Darwin)
1867				*Das Kapital* (Karl Marx)
1871				Commune de Paris, 1871
1874	Enrolled at University of Edinburgh; enrolled at Royal School of Mines, London			
1877	Demonstrator in practical physiology, University College, London	Visited *Paris, Italy, Mexico*		
1880	Lecturer in Zoology, University of Edinburgh			
1881			'The classification of statistics and its results'	
1885		European travel (to 1900)		
1886	Married Anna Morton			
1887		James Court Project, Edinburgh; first Summer School convened, Edinburgh		
1888	Application for chair of botany, University of Edinburgh, rejected			
1889	Appointed professor of botany, University College, Dundee			
1890s			*The evolution of sex* (with J. Arthur Thomson)	
1892		Outlook Tower, Edinburgh opened		
1893			*Chapters in modern biology*	
1897		Visit to Cyprus		
1898			'The influence of geographical conditions on social development'	

Dates	Life and career	Activities, travel, fieldwork	Publications	Contemporary events and publications
1900	Formed publishing firm 'P.G. and Colleagues'	Great globe project, Paris (with Paul Reclus) At Paris Exposition	*Man and his environment*	Paris Exposition
1902		British National Institute of Geography project	'Edinburgh and its region'	
1903		Civic survey of Dunfermline		
1904			*City development*	
1906			*First visit to Outlook Tower*	
1907			(in association with Victor Branford) *The making of the future* series initiated	
1910		First Cities Exhibition, Burlington House, London		
1912	Refused knighthood		*The masque of learning*	
1913		Congress of Cities, Ghent		
1914				Outbreak of war
1915			*Cities in evolution*	
1916		First visit to India Town Planning Reports, India and Ceylon (until 1919)		
1917	Deaths of Anna Geddes and Alasdair Geddes (elder son)			
1919	Retired from Chair of Botany, Dundee	Consultant for Hebrew University, Tel-Aviv		Treaty of Versailles
1920	Professor of Civics and Sociology, University of Bombay		'Essentials of sociology in relation to economics'	
1924	Founded Scots College, Montpellier; settled in France	Development of regional survey courses at Montpellier		
1925			'Talks from my Outlook Tower'	
1928	Married Lillian Brown			
1929			'Rural and urban thought'	Wall Street crash

Dates	Life and career	Activities, travel, fieldwork	Publications	Contemporary events and publications
1930s				World depression
1931			*Life: outlines of biology*	
1932	Knighted; died at Montpellier 17 April			

Geoffrey Edward Hutchings 1900-1964

Photograph by courtesy of John Sankey

KEITH WHEELER

1. EDUCATION, LIFE AND WORK

Geoffrey Edward Hutchings, born on 12 December 1900, was the second son of a builder and amateur geologist living at Strood, near Rochester, Kent. At the age of sixteen he left the Rochester Technical Junior School and began a five-year engineering apprenticeship at the Royal Dockyard, Chatham. He qualified in 1921, and then served as an engineering draughtsman for a further five years.

His early scientific interest was in geology. He joined the Rochester and District Naturalist Society, and his first publication was a Geological Report in the Society's journal, *The Rochester Naturalist*, 1922-3. At a meeting of the Society he became acquainted with C.C. Fagg, an exponent of Sir Patrick Geddes's methods of regional surveying and an outstanding amateur field scientist. Henceforth Hutchings was converted to the educational and scientific objectives of regional surveying as a framework for his own field studies which also coincided with his increasing interest in teaching.

Between 1921 and 1926 he taught science and engineering to evening classes at Medway Technical College and studied at Birkbeck College. Here he attended, among other classes, the lectures of the eminent London geologist, G.M. Davies. Later he studied botany and zoology but he did not proceed to a degree. It was about this time, too, that he met S.W. Wooldridge (1900-63), Professor of Geography at Birkbeck College from 1944-7 and then at King's College. Both Fagg and Wooldridge were to remain lifelong friends and collaborators of Hutchings and they profoundly influenced his career and geographical thought. In 1926 he returned

to the Rochester Technical Junior School as a full-time teacher where he experimented with the teaching of biogeographical field studies at a time when science in boys' schools was mainly confined to physics and chemistry.

He continued to apply himself to regional surveying and acquired a detailed topographical knowledge of Kent, and other parts of the South East. As Honorary Secretary of the Regional Survey Section of the South East Union of Scientific Societies he edited or wrote papers describing the results of regional surveys or investigations into physical geography. In 1928 he contributed chapters on topography to two Kent Regional Planning Schemes; and he also collaborated with C.C. Fagg in writing the chapter on 'The South East' in *Great Britain: essays in regional geography*, edited by A.G. Ogilvie; this work for many years became a standard university regional textbook. The other contributors were university geographers, but Fagg and Hutchings were invited because of their regional survey work. In 1930 they collaborated again, and wrote *Introduction to regional surveying* as a manual for teachers and students in these methods. This book, along with the publications of the Le Play Society, did much to introduce geographical fieldwork into the curricula of schools and colleges.

Through his association with the Misses Pugh at their Centre for Rural Education, Hill Farm, Stockbury, Kent, Hutchings was able to develop his approach to field study as an educational and interpretative activity rather than one of scientific research. He described, in collaboration with Christine Pugh, the Centre's surrounding topography in a privately published book, *Stockbury: a regional study in North East Kent* (1928). The use of Hill Farm as a training place for teachers and students in regional survey was explained in a manuscript prepared by Hutchings in 1937 with the intention of getting it adopted as a Field Study Centre by a local education authority or the University of London.

Between the years 1938 and 1945 he was in the Middle East serving as Principal of the Technical School first in Baghdad, Iraq, and then on the Persian Gulf Island of Bahrain. In January 1943 he wrote to F.H.C. Butler with whom he had worked before the War on the Executive Council of the School Nature Study Union, and enquired about the possibility of establishing field study centres as part of the envisaged post-war reconstruction programme in education. Hill Farm was referred to as an example of the kind of Centre that might be set up. He had discussed this possibility with Butler before leaving for the Middle East. This letter may have stimulated Butler into greater action though it was not until December 1943 that he called the inaugural meeting of the Council for the Promotion of Field Studies (now the Field Studies Council). In the same year Hutchings was also corresponding with C.C. Fagg who became the Council's first Treasurer. From the beginning Hutchings was recognized as a strong candidate for the Wardenship of one of the Field Study Centres that the CPFS began to establish. In 1946 the National Trust leased Juniper Hall, near Dorking, Surrey to the CPFS, and Fagg moved in as Acting Honorary Warden until Hutchings was able to take up the full-time post. This he did in 1947 having returned from the Middle East in 1945 and then having worked for the Ministry of Town and Country Planning in East Anglia for two years.

From now on Hutchings identified himself more closely with geography, and his concept of educational field studies was strongly influenced by Professor S.W. Wooldridge, Chairman of the Juniper Hall Committee, and later an important member of the CPFS Council. They both conceived of geographical field work as a method of acquiring an 'eye for country' through scientifically observing the countryside and learning to interpret the evolution of landscape. To this process Hutchings contributed his geographical landscape drawing technique derived from his early training as a draughtsman, and his innate ability as an artist. In 1955 he issued from Juniper Hall Field Centre a printed booklet, *An introduction to geographical landscape drawing*, which proved successful enough for him to expand it into a book, *Landscape drawing* (1960) with an introduction by D.L. Linton (1906-71), Professor of Geography at the University of Birmingham, and also an exponent of field sketching. This book combined the exactitude of observational science with the interpretative power of an artist, and may be regarded as a classic of its kind. Perhaps more than any other publication arising from the work of the CPFS it summed up the twin aims of the Council in its efforts to encourage field studies in the widest sense having both scientific and aesthetic objectives. In particular it typified Hutchings's own interdisciplinary interests which were those of a gifted amateur who could express his scientific observations through painting, drawing, and precise prose. Indeed, most of the publications he was involved in were illustrated by his maps and drawings.

The Juniper Hall programmes produced by Geoffrey Hutchings were notable for the inclusion of courses for non-academic pupils from Secondary Modern Schools. He was Warden for nine strenuous years, and he also participated fully in the affairs of the CPFS. In 1956, partly as a result of ill health, he retired from the post, and was appointed CPFS Senior Tutor in Geography. This enabled him to travel between the several field centres and advise the less experienced Wardens. The culmination of his career was marked by election as President of the Geographical Association in recognition of his services to geographical education. In addition, he served on the Council of the Royal Geographical Society (1961-4) and became Chairman of the Education Committee in June 1963.

Hutchings was married in 1923 and had two sons. In appearance he has been likened to a distinguished Spanish nobleman with his neatly trimmed triangular beard and long head. He is remembered as a patient and gifted teacher with a quiet sense of humour who brought meticulous craftsmanship to everything he did. The wise advice he gave to the other Field Study Wardens is recalled with affection and respect. He died unexpectedly on 21 February 1964 after a short illness. His last publication, an account of a fieldwork excursion by S.W. Wooldridge who had died the previous year, appeared in *Guide to London excursions* (1964).

2. SCIENTIFIC IDEAS AND GEOGRAPHICAL THOUGHT

Hutchings described his geographical philosophy in an address to the School Nature Study Union, 'The geographer as field naturalist' (1949). He argued that the insight of the natural historian combines with that of the physical geographer to explain the inter-relationships expressed in the regional division of the earth's surface and the appearance of landscape. For him, the study of landscape emphasized the ideas of geography as a field science, and he was critical of what he called 'paper geographers' who culled their facts from documents only. Essentially he regarded himself as a teacher whose aim was to extend the students' visual experience through practice in interpreting what they saw, and by giving them the best possible 'picture' of a piece of country. Undoubtedly, he was in accord with the dictum of W.M. Davis, the American geomorphologist and Wooldridge's mentor: 'Fieldwork involves four imperatives: go, see, think and draw'.

Hutchings was not overtly concerned with the world view of geography. However, arising from his lifelong interest in biogeography (he was said to be better at interpreting the botanical landscape than Wooldridge), he made in his Presidential Address in 1962 to the Geographical Association a perceptive statement of more than local interest when he declared:

> No provision of geographical field teaching is complete without provision for the study of plant and animal ecology... Who knows that it will not one day become the responsibility of properly informed geographers rather than anyone else, to point out mistakes and redirect the application of much specialised knowledge in a campaign to

secure the very survival of mankind on the earth? If the responsibility for this does eventually fall to geographers the teachers of this generation should anticipate it by giving the subject of biogeography a prominent place in their teaching, especially that part of their teaching which is done in the field, the principal source of inspiration of the study of man in the realm of nature.

The relevance of his point of view to the present day concern about the ecological environment shows considerable foresight.

3. INFLUENCE AND SPREAD OF IDEAS

Despite the value he placed on biogeographical studies Hutchings published very little on this subject. In 1959 he recommended in an address to the Union of Local Scientific Societies that they should carry out ecological mapping surveys of their region. His main influence, apart from field sketching, was through the codification and exemplification of fieldwork methods illustrating rural regional interrelationships set in the scarplands of south-east England. In 1956 he collaborated with S.W. Wooldridge in producing *London's countryside*, which became a model for writing up fieldwork excursions in rural areas and was adopted by numerous geographical publications subsequently.

His work first at Hill Farm and then at Juniper Hall was a major pioneering contribution to the development of field study centres in country areas set up originally by the CPFS, and now copied by many other organizations here and abroad. As D.L. Linton noted, Hutchings more than anybody else created the role of Warden. But it was his demonstration of the value of landscape drawing (or 'geographical draughtsmanship' as he called it) which was his unique contribution to fieldwork methodology. Recently, too, W.G.V. Balchin has identified Hutchings's sketching methods as an important contribution to the acquisition of the basic skill of geography: graphicacy. Thus, although Hutchings was a masterly exponent of observational geography, it was as a teacher of field studies that he excelled, and it is in his contribution to the advance of geographical education in Britain that his major influence continues to be felt.

Bibliography and Sources

1. OBITUARIES AND REFERENCES ON G.E. HUTCHINGS

Obituary by F.H.C. Butler and Ian Mercer, *Annual Report Field Studies Council*, 1963, 3-5

Obituary by F.H.C. Butler, *Natural Science in Schools*, vol 2, (1964), 38-9

Obituary by David L. Linton, *Geography*, vol 49, (1964) 135-6

Obituary, *Geogr. J.*, vol 130, (1964), 319-20

W.G.V. Balchin. 'Graphicacy', *Geography*, vol 57, (1972), 185-95. This article points out the importance of Hutchings's method of landscape drawings as a contribution to graphicacy.

2. WORKS BY G.E. HUTCHINGS

1928 (with C.C. Fagg) 'The South East' with a contribution by Professor A.G. Tansley, 19-41 in *Great Britain: essays in regional geography* by Twenty-six authors... Ed Alan G. Ogilvie. Published on the occasion of the 12th International Geographical Congress at Cambridge, 1928, 502p.

---- (with C. Pugh) *Stockbury: a regional study in North-East Kent*. The Hill Farm, Stockbury, Kent, Rochester, 1928, 71p.

1930 (with C.C. Fagg) *An introduction to regional surveying*, Cambridge, 1930, 150p.

1949 'The geographer as field naturalist', *School Nature Study*, vol 44, (1949), 33-8

1957 (with S.W. Wooldridge) *London's countryside* (Geographical Field Work for Students and Teachers of Geography), London, 1957, 223p.

1960 *Landscape drawing*, London, 1960, 134p.

1962 'Geographical field teaching', *Geography*, vol 47, (1962), 1-14

Annual Reports of the CPFS/FSC 1948-56 contain Hutchings's yearly Juniper Hall Field Study Centre Warden's Report

3. UNPUBLISHED SOURCES

The following, with other documents concerning the early years of the CPFS, are held by the School of Education Library, University of Leicester. This archival material also contains a complete bibliography compiled by Keith Wheeler.

The Hill Farm, Stockbury, Kent. A rural educational centre. Draft descriptions by G.E. Hutchings, corrected by Miss T.M. Pugh, Sept. 1937

Letters to and from F.H.C. Butler and C.C. Fagg during the years 1943 and 1944

Keith Wheeler is Senior Lecturer in Geography at the City of Leicester College of Education.

Dates	Life and career	Activities, travel, fieldwork	Publications	Contemporary events
1900	Born near Rochester, Kent, 12 December			
1916	Engineering apprenticeship Royal Dockyard, Chatham			
1919				End of First World War
1921	Qualified as engineering draughtsman employed at Chatham, Woolwich and the Admiralty; teaching evening classes at Medway Technical College and studying at Birkbeck College	Member of Geological Association and an active member of Rochester and District National History Society; involved as weekend tutor at Hill Farm, Rural Education Centre Stockbury, Kent		
1926	Left engineering and took up full-time teaching at Rochester Junior Technical School			
1927		Elected Fellow of Geological Society		
1928			(with C.C. Fagg) 'The south east' in Great Britain: *Essays in regional geography*	
1930			(with C.C. Fagg) *Introduction to regional surveying*	Le Play Society formed to promote regional surveying
1938	Principal of Technical Schools in Iraq and Island of Bahrain	Resident in the Middle East		
1939				Outbreak of Second World War
1943				Council for Promotion of Field Studies founded by F.H.C. Butler (later known as Field Studies Council)
1945	Returned to England; working in Ministry of Town and Country Planning	Elected Fellow of the Royal Geographical Society		End of Second World War
1946				First CPFS Field Study Centre opened at Flatford Mill, Suffolk
1947	Warden of Juniper Hall Field Study Centre, Surrey			

Dates	Life and career	Activities, travel, fieldwork	Publications	Contemporary events
1949			'The geographer as field naturalist'	
1956	Left Wardenship and became FSC Senior Tutor in Geography			
1957			(with S.W. Wooldridge) *London's countryside*	
1960			*Landscape drawing*	Le Play Society wound up
1961		President, Geographical Association 1961-2 Member of Council and Chairman of Education Committee of Royal Geographical Society 1961-4		
1962			'Geographical field teaching'	
1964	Died 21 February 1964			Seven FSC Field Study Centres open

Bartholomäus Keckermann
1572-1609

Contemporary engraving by an unknown artist

MANFRED BUTTNER

As a scholar Bartholomäus Keckermann ranks among the greatest thinkers of his time. An intellectual leader of note in the German Reformed (*Deutsch-Reformierten*) Church, he gave a new direction to scientific thought with his analytical-distinctive method. His major contribution to Geography was given in his *Systema Geographicum*, from which Varenius took over whole paragraphs without acknowledgement. This influenced geographical thought for several generations, indeed to the present time for the division into *geographia generalis* (general geography) and *geographia specialis* (area studies) can be traced back to Keckermann. He redefined geographical aims, functions and methods: he also separated geography clearly from other disciplines, notably from theology, as no previous writer had done and he could therefore be described as the first scientific theoretician of modern European geography and the founder of post-Reformation scientific geography.

1. EDUCATION, LIFE AND WORK

Bartholomäus Keckermann was born and spent his childhood in the Free City of Danzig. His birth is generally dated as 1572 though Zuylen has said that 1571 or 1573 are also possible dates. His father, Georgius (or Gregor) Keckermann, came from Stargard in Pomerania, where he had lived at the court of Barnim, and settled in Danzig, first as *Konrektor* of the Marienschule and then as a merchant. He married Gertruda Lorengi, a woman of great integrity and firm religious principles. Gregor Keckermann's move from Pomerania appears to have been encouraged by his brother Joachim, who was pastor of the Church of St. Johannes in the eastern half of Danzig. Bartholomäus was taught at home for some time but later attended the Danzig *Gymnasium*, of which the head was J. Fabricius, a convinced Calvinist. At this time tension was growing between Calvinists and Lutherans and this eventually caused Joachim Keckermann to leave Danzig in 1588. Apparently the young Bartholomäus accompanied his uncle to the famous University of Wittenberg. Many Danzigers were then registered at Wittenberg (where Joachim Keckermann had been a student) and Calvinism had been firmly established in the University after Luther's death. Even the then elderly Melanchthon was said to be a crypto-Calvinist.

At Wittenberg Bartholomäus Keckermann read philosophy and theology and of his teachers the philosopher Claepius had the strongest influence for with others he turned the student's mind to Aristotelian thought. Looking back later Bartholomäus Keckermann described his earlier way of thinking as 'aimless wandering through the sandy and all but barren realms of Ramist philosophy'. (Ramism favoured reason rather than authority and attacked scholasticism.) Having accepted Aristotelian views, Keckermann derived stimulus from the philosophical, including the geographical, work of Melanchthon.

While at Wittenberg Bartholomäus visited Leipzig, where Ramism had already been abandoned. At Leipzig his Aristotelian approach received further encouragement in the lectures of Neldelius. Keckermann also visited the University of Altdorf: though he never studied there he was an occasional guest of Scherbius, whose work he knew well, and also corresponded with

him and with a number of other friends, some of whom
sent texts of their lectures.   After four years at
Wittenberg he went to Heidelberg, renowned for the
Heidelberg Catechism, the 'bastion of the German
Reformierten'.   He matriculated at Heidelberg on
22 October 1592 and left the University in 1602, after
a long and crucial study of theology and philosophy.
He took the degree of *Magister Artium* followed in 1602
by the *Licentiatus Theologiae*.   This long period of
education he conceived as a preparation in philosophical and theological disciplines for service in
education and the church.

Frederick IV, Elector of the Palatinate, followed
the enlightened tradition of appointing foreigners as
preachers and teachers from time to time and Keckermann
became one of the four Regents of the Bursary Students'
Residence.   His duties were to teach some of the
subjects of the Faculty of Philosophy, Dialectics,
Rhetoric and Grammar.   He also taught the rudiments
of science at the *Pedagogium* from which many of the
students were going forward to the Sapienzkolleg, a
residential theological college.   His degree of
Magister Artium qualified him to lecture in logic in
the Faculty of Philosophy and in February·1600 he received permission to teach Hebrew also.   At this time
he also lectured on Dogma.

In 1601 Keckermann stayed at Zurich during September and October and it was probably there that he met
the group of authors compiling Sebastian Munster's
Cosmography, a work which appeared in corrected and
enlarged editions until the middle of the seventeenth
century.   This work is critically discussed in Keckermann's *Systema Geographicum* of 1602.   On his return
from Switzerland in 1601 Keckermann was invited to
accept an appointment at the *Gymnasium* in Danzig which
he accepted, having refused a similar offer four years
earlier on the ground that his studies were incomplete.
The choice was not easy, for Keckermann loved Heidelberg though he regarded the Danzig offer as an indication of the divine purpose he must serve by defending
the truth in his native city and Fatherland against
the Counter Reformation.   Though he was offered a
chair of theology in Heidelberg, his choice was firm.

Had he stayed in Heidelberg as a theologian he
might never have written any geography at all but at
the Danzig *Gymnasium* he was so shocked by the lack of
any good geography text that he decided to write one
himself.   The teaching was almost of university standard and courses were attended by students from various
universities.   Among them were young noblemen from
Poland, Livonia and Courland: the *Gymnasium* had the
reputation of possessing more learning than any comparable institution in Prussia, Livonia, Courland and
Further Pomerania.   It was also renowned as a bulwark
of Protestant Christianity in north east Germany;
in 1605 all the teachers were Protestants and the
city council of Danzig was determined that it should
remain so.

Though he became the *Konrektor* of the *Gymnasium*,
Keckermann refused to carry out any administrative
duties and concentrated entirely on teaching and research.   His teaching was arranged on a three-year
cycle, with logic and physics in the first year, meta-
physics and mathematics (including geography) in the
second and ethics, economics and politics in the third.
The last seven years of his short life were spent in

this way and all the four hundred students of the
Gymnasium followed his courses.   Among them was
Goclenius and through him, and many more of less renown, Keckermann's influence was radiated.   It was
also spread through his writings.

Keckermann was both a philosopher and a theologian
though his main strength lay in philosophy, and at
Heidelberg he wrote mostly on logic, but also on dogma.
At Danzig he wrote many short essays on philosophical
subjects and also some comprehensive works on meta-
physics, physics, geography, astronomy, rhetoric and
history.   His major work was the *Philosophia Practica*,
which concerned ethics, economics and politics.   It is
probable that the earthquake of 1601, which he experienced in Switzerland, strengthened his interest in
geography and some of his letters, recently found in
Zurich but not yet investigated, may confirm this view.

On Keckermann's death his student Alsted published
several of his writings, almost all derived from his
lectures.   Some of the publications went through
several editions, especially those on geography but it
is not always possible to say exactly when they were
written, though some evidence is given by reference
to events in his lifetime or earlier.   For example
Keckermann once mentioned the discovery of America a
hundred and ten years earlier, which would give the
date 1602.   Editions of his work described as complete appeared posthumously as *Systema Systematum*,
Hanau, 1613, and *Opera Omnia*, Geneva, 1614, but in fact
neither is complete for various other writings were
discovered later.

Melanchthon was probably the major influence on
Keckermann, and led him to an Aristotelian view of
philosophy and geography.   The works of Claepius and
Scherbius also led him to Aristotle who, Keckermann
said, gave him an ordered and comprehensive methodology sharply contrasted with the confused thought
of Plato.   It may be that Keckermann derived his
Aristotelian outlook directly from Melanchthon but
possibly he found that his own views were confirmed
by the argument of Melanchthon.   Whatever the mental fructification, the practical result was that
Aristotelian science dominated the teaching in
Protestant schools, colleges and universities.   But
this was not the only source of Keckermann's philosophy
and theology, for he was conversant with the humanist
writing of the time.   Significantly, he disagreed
completely with Calvin's conviction that study of the
natural sciences led people from the knowledge of God
to the knowledge of sin.

2. SCIENTIFIC IDEAS AND GEOGRAPHICAL THOUGHT

More is known about Keckermann as a theologian and
philosopher than as a geographer, though his *Systema
Geographicum* of 1611 gave a new approach to the subject.   The *Geographia Generalis* is the major part of
the *Systema*, in 163 pages followed by an appendix of
31 pages.   There are numerous references to earlier
geographical writings by Mercator, Apian, Ortelius,
Münster, Eratosthenes, Pliny the Younger, Aristotle
and others and all quotations are set in italics with
the title of the book and the chapter as a reference.
This was not a universal practice of the time -- or
indeed of a later time -- but Keckermann had learned
its value in his theological studies.

The *Systema* deals first with the essential scientific basis of geography, the science of the measurement of the earth's sub-division. The whole earth, and not merely the inhabited areas, as Ptolemy stated, must be studied but geography dealt only with land and water and as such was part, but only part, of cosmography which also included physics and astronomy. And geography was initially divisible into two parts, *generalis* and *specialis*.

General geography as a concept needs sub-division. There is the *absoluta* theory, the basic whole, and the *comparata*, the comparison of different areas. Geography must deal with real things, *realis*, and with their transfer to a map, that is their pictorial representation, *pictoralis*. Geography may be general, theoretical, real in unity and as such will have certain qualities (*affectiones generales*). These were treated by Keckermann under five headings. Of these the first is land and water, a single entity, for the geographer of interest for their quantities, dimensions and distances but for the physicist for their dampness, dryness, warmth, cold and other attributes. Secondly, land and water form a spherical body, as many earlier geographers had thought, and therefore there was one centre of the globe and only one centre of gravity. Thirdly, the existence of high mountains and deep seas does not detract from the sphericity of the earth. Fourth, the earth is the immovable centre of the universe, as Aristotle showed in his theory of the elements. Fifth, the earth is only a small point in comparison with the entire universe, as Aristotle, Ptolemy and Pliny the Younger agreed.

Figure 1.
Keckermann's Principle of the Division of Geography

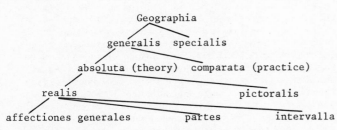

Figure 1 illustrates this general approach. Having established the five principles, Keckermann then considers more detailed aspects, working from left to right on the bottom line of fig. 1. The *affectiones generales* is followed by the *partes*, the *affectiones partiales*. This is descriptive rather than explanatory and begins with the *principalis*, the division of land by water, using the terms, *continens, insula, peninsula, isthmus, promontium, mons, vallis, campus, sylva*. Keckermann regarded a mountain as a high pile of earth and refers to known high mountain ranges without discussing their origins, though he thought that they must have existed before the flood. Possibly some mountains were formed after the flood when great quantities of earth were lifted up by earthquakes. After this theoretical statement, Keckermann gives a thorough treatment of mountains drawn from the works of Pliny the Younger, Solinus, Strabo, Mela, Nonius, Sues, Cardanus, Eratosthenes and others. He also describes

the appearance of *vallus, campus* and *sylva* but with no comment on their origins.

Figure 2.
Keckermann's System of 'Morphological' Terms

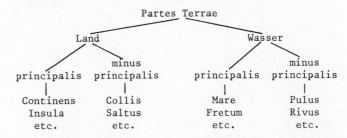

There naturally follows a description of lesser *principalis*, the *collis, saltus, arbustum, virgultum, demetum* (Fig. 2). After the land features those of water are discussed under *partes principales* with *mare, fretum, sinus, stagnum, flumen* and under *minus principales pulus, rivus, lacuna, cisterna, piscina*. There is no new geographical material here but rather a scheme for systematizing study of known features.

Measurement in Keckermann's mind was based on the human body, and he quoted with approval Plato's dictum that man is the measure of all things. The *minora mensurae* are based on the *granum*, four of which form one *digitus*, of which in turn, four form one *palmus* and finally four give one *pes*. The 'foot' therefore equals four *palmi*, sixteen *digiti* and sixty-four *grani*. The major measures consist of the *passus geometricus*, of five *pedes*, the *stadium* of twenty-five *passus* and the *milliare* of one thousand *passus*. In Keckermann's time there were several local or national systems of measurement and on his plan the distance between the meridians at the equator was fifteen *millaria* and the circumference of the earth 5,400 German miles. The whole earth had a division by lines of latitude and longitude, based on the equator, which remained fixed while the horizon was of variable extent. The central meridian of the longitude went through the Canary islands.

Having established the mathematical form of the globe, Keckermann then dealt with the second part of the *distinctio particularis*, the *orta*, which was subdivided into the *distinctio zonalis* and the *distinctio climatis*. The former was treated purely mathematically and not regionally, for he appeared to be little concerned with the problems of the antipodes, the inhospitable character of regions scorched by sun, or the influence of excessive heat on people. That certain regions are particularly suitable for human habitation is mentioned only briefly in a footnote. He divided the climates of both inhabited and uninhabited areas between the equator and the Arctic circle (which he thought to be at 66° 31') into twenty-four types, of which all except the seven most northerly in Norway could be located by some known area or place. These places (from south to north) included Rhodes, Rome, Venice, Podolia, Wittenberg, Rostock, Ireland, Riga, Gothia, Bergen, Viburg, Scotland and Dalecarlia.

The basic information so far discussed was the 'real' part of the *geographia generalis absoluta* and Keckermann then turned to the *geographia generalis*

absoluta pictoralis.   This included the study of
globes and maps.   He mentions the famous map collections of Mercator and Ortelius and discusses the maps
of the entire earth's surface in these and other
collections.   Ptolemy mapped only parts of the earth's
surface and therefore his maps were *specialis* but since
Münster produced maps of the whole earth in the early
sixteenth century there had been universal maps, classified as *generalis*.   These were geographically based
on greater and less circles, the former equatorial and
the latter meridional: there were also nautical
maps for navigation.   Keckermann realized that some
maps were of general interest and some had a directly
practical use, for example by travellers on land
and sea.   Each offered problems of compilation and
Keckermann's pragmatic approach to the eventual use of
the map was itself a pioneer scientific contribution
to cartography.

Having discussed the *geographia generalis absoluta
pictoralis* (Fig. 1), Keckermann turned to the *geographia generalis comparata*.   Characteristically he
took a strongly mathematical approach in marked contrast to the theologically based 'anthropogeography' of
Münster and Neander.   He recognized the significance
of length of daylight to people in various parts of the
earth and with that the varying length of shadow but
did not follow this with a study of the inhabitants or
of men and environment.   His main interest lay in
latitude and longitude and he argued that there was no
theological objection to the existence of people in
antipodean areas.   As a teacher he was concerned to
give his Danzig students a firm sense of relative location on the earth's surface.

The *geographia specialis* could be divided into the
realis and the *pictoralis*.   Each part of the known
world is treated under its anthropogeographical and
physiogeographical aspects.   Inevitably there is a
division between the Old and New Worlds, and the continents of Europe, Asia and Africa should be treated
before America.   For Europe he gives the mathematical
co-ordinates with some cultural and physiographic
material.   He deals with European culture as a unitary
whole and then enumerates the empires, kingdoms and
states.   He makes no use of reports of travellers
and others and in fact merely gives a descriptive
treatment.   The physiographical content lists such
phenomena as rivers, lakes and mountains.   On Asia he
gives a similar survey, but beyond mentioning its
significance for Christianity there is no trace of a
geographia sacra.   The account of America is preceded
by a short history of its discovery and he gives a list
of maps available for study.   For this, as for his
basic thought on area studies, he draws material from
Ptolemy, Münster, Mercator, Neander and Chynäus.

Although Keckermann's geographical statements are
like those of his predecessors, especially Münster,
Mercator, Peucer and Neander, his geography is in no
way subsidiary to theology.   In this he was partly
foreshadowed by Mercator but his originality emerges
in five ways.   First, he offered a new terminology
as a means of analysis for a growing mass of material.
Second, he drew attention to location, especially in
Germany, and particularly for climate.   Third, he
established the apartness of the geographer from the
physicist, as one researching on the entire globe with
its land and water masses.   Fourth, he showed a wide
knowledge of the significant geographical work of his
time, both in Germany and beyond it for he quoted the
Italian scholars Patritius, Gauricus and Cardanus, the
English Camden, the French Bodin and the naturalized
Chynäus and the Spanish Acosta.   Fifth -- and most
important -- he refused to be bound by the chronology
of the Book of Genesis.   In this he differed from his
immediate predecessors such as Münster and Mercator
who had presented the geographical material derived
from Aristotle, Ptolemy and others with the Book of
Genesis as a base.   He also rejected the 'external to
internal' order with fire, air, water and earth studied
in sequence as this would lead to physics or cosmography rather than geography.   Having discarded the
Aristotelian theory of the elements, followed by
Reisch, Glarean, Apian and Melanchthon, Keckermann
became the first scientific theorist of geography.

Keckermann based his separation of the natural
sciences from theology on his interpretation of the
story of the Fall of Man.   First, he recognized the
division of all knowledge into theory and practice but
he added that theoretical discussion had no value in
itself: therefore all theology must be primarily concerned with its essential purpose, the salvation of
man.   Second, theological analysis must be based on
human valuation and this gave it distinction from all
other subjects.   Third, as a theologian Keckermann
agreed that the human aim was to regain the image of
God, lost at the Fall, for only so could the individual, and indeed mankind, achieve salvation.   The
likeness of man to God could be regained in two ways,
either through divine revelation to the theologian or
by study of philosophy (which included the natural
sciences and geography).   Fourth, because Keckermann
thought that God was everywhere and in everything,
knowledge was itself a way to God.   The deeper man's
knowledge and use of the earth, the more he would regain the former image and likeness of God lost at the
Fall.   In the earlier state of paradise, Man knew
more about nature than all the sinful scholars of a
later time but learning could bring a restoration to
the earlier state of perfect knowledge.   Fifth, as
the Fall destroyed mankind's subjective knowledge, a
fragment of God's image and likeness, the *Imago Dei*,
remained in the objective area of natural knowledge,
and therefore the less science examined mankind itself
and its subjective relationship with God, and concentrated on natural phenomena, the sooner it could help
mankind to regain the divine image and with it the
human power over nature.   Man can only attain the full
image and likeness of God, however, by following at one
and the same time the scholarly paths of philosophy
and theology.   Sixth, Keckermann rejected 'natural
theology' for knowledge of the natural sciences did not
lead mankind to knowledge of God through the revelation
of divine providence (as Melanchthon believed) but
rather gave man power to become like God.

This purpose could be best served, Keckermann believed, by giving full liberty to geography, physics
and other sciences to develop their own scientific
activities and research methods.   A compromise between
the natural sciences and *Doctrina Evangelica*, practised
by Melanchthon, or between classical geographical
ideas on the one hand and biblical ideas on the other,
which Mercator was still trying to achieve, should not
even be attempted.   To Keckermann the liberation of

geography from theological control was itself a vital contribution to theology and he was perhaps fortunate not to experience the controversies on science and religion of the eighteenth and nineteenth centuries.

3. INFLUENCE AND SPREAD OF IDEAS

Even the opponents of Keckermann, such as Martini, recognized his authority as a scholar and his work on logic was widely studied. Pareus wrote that Keckermann's works achieved recognition throughout Europe in a very short time. Alsted gave particular praise to his *Systema Geographicum* as a new basis for philosophical thought. Later geographers such as Göllnitz, Christiani and Varenius not only adopted Keckermann's system but reproduced whole passages from his major work. Undoubtedly the main contribution of Keckermann was the general and special (*generalis, specialis*) approach which made it possible to arrange the geographical knowledge of the time in an orderly manner. And the separation from other subjects, notably from theology, gave it identity as a scientific discipline, with its own technique of analysis. Also of value, but not always followed by later writers, was his insistence on giving sources in using the work of others.

Though Keckermann was regarded as a major geographer for a long time after his death, by 1700 Varenius was considered by many to be the originator of the *geographia generalis*. This is partly because Varenius published a *Geographia Generalis* in 1650 in Amsterdam which was admired by the physicist Sir Isaac Newton so much that he edited two Latin editions in 1672 and 1681. Varenius had studied Newtonian physics and his geographical ideas were partly acquired from Nathaniel Carpenter, of Exeter College, Oxford, whose *Geography delineated in two books* of 1625 and 1635 was partly based on Keckermann's *Systema*. The strength of Keckermann as a scholar lay in his breadth of learning with his vision of the relation between science and religion. Throughout his working life as a leader of the German Reformed (*Deutsche Reformierten*) Church he was a student of Calvinist theology but he never accepted all the views of its founder John Calvin. Above all, he saw a future for geographical study on a world as well as on a local scale and many of his ideas were incorporated in the works of later writers, as well as in the teaching of geography from his time onwards.

Bibliography and Sources

1. REFERENCES ON BARTHOLOMAUS KECKERMANN

a. General

Holtzmann, 'Bartholomäus Keckermann', *Allg. Dtsch. Biogr.*, vol 15, Leipzig, (1882), 518

Petersen, P. *Geschichte der aristotelischen Philosophie in Deutschland (History of the Aristotelian philosophy in Germany)*, Leipzig, (1921), 542p. *passim*

Zuylen, W.H. van *Bartholomäus Keckermann, Sein Leben und Wirken (Bartholomäus Keckermann, his life and work)*, Theologische Inauguraldissertation Tübingen, Borna-Leipzig, (1934), 190p.

Schieder, 'Bartholomäus Keckermann' *Altpreussische Biogr.*, vol 1, Königsberg, (1941), 329

Heppe-Bizer, *Die Dogmatik der evangelisch-reformierten Kirche (Dogmatics of the Protestant-Reformed Church)*, Neukirchen, (1958), 584p. *passim*

Nadolski, B. 'Barttomiej Keckermann', *Pol. Słownik Biogr.*, vol 10, Wroclaw-Warszawa-Kraköw, (1966-7), 322-3

b. As a Geographer

Philippson, A. 'Two forerunners of Varenius', *Ausl.*, (1891), no 52, 817-18

Büttner, M. 'Theologie und Klimatologie' (Theology and climatology), *Neue Z. fur syst. Theologie und Religionsphilos.*, vol 6, (1964), 154-91

Büttner, M. 'Geographia Generalis before Varenius', *International Geography*, 1972, I.G.U. Congress, 948-50

Kastrop, R. *Ideen über die Geographie und Ansatzpunkte für die moderne Geographei bei Varenius unter Berücksichtigung der Abhängigkeit das Varenius von der Vorstellungen seiner Zeit (Varenius' ideas on geography and their modern relevance, with his use of current concepts)*, D. Phil. diss., Universität des Saarlandes Saarbrucken, 1972, 210p.

Büttner, M. *Die Geographia Generalis vor Varenius, Geographisches Weltbild und Providentialehre (The Geographia Generalis before Varenius, geographic conception of the world and doctrine of providentia)*, *Erdwiss. Forsch.*, vol 7, Wiesbaden, (1973), 251p.

Büttner, M. 'Keckermann und die Befründung der allgemeinen Geographie. Das Werden der Geographia Generalis im Zusammenhang der wechselseitigen Beziehungen zwischen Geographie und Theologie' (Keckermann and the foundation of Geographia Generalis. The development of Geographia Generalis in connection with the reciprocal relationships between geography and theology), *Plewe-Festschrift*, Wiesbaden, (1973), 63-9

Büttner, M. 'Kant und die Uberwindung der physiko-theologischen Betrachtung der geographisch - kosmologischen Fakten (Kant's explanation of physicotheological aspects of geographico-cosmological facts)', *Erdkd.*, vol 29, (1975), 53-60

Büttner, M. 'Kant and the physico-theological consideration of the geographical facts', *Organon*, Warsaw, vol 11, (1975), 233-50

Büttner, M. 'Die Neuausrichtung der Geographie im 17 Jahrhundert durch Bartholomäus Keckermann, Ein Beitrang zur Geschichte der Geographie in ihren Beziehungen zu Theologie und Philosophie' (The new orientation of geography by Bartholomäus Keckermann in the 17th century. A contribution to the history of geographical thought with respect to its relationships to theology and philosophy), *Geogr. Z.*, vol 63, (1975), 1-12

Büttner, M. 'Die Emanzipation der Geographie zu Beginn des 17. Jahrhunderts. Ein Beitrag zur Geschichte der Naturwissenschaft in ihren Beziehungen zur Theologie' (The emancipation of geography at the

beginning of the 17th century. A contribution to
the history of natural science and its relationship to theology), *Sudhoffs Arch.*, vol 59, (1975),
148–64

Büttner, M. 'Die Neuausrichtung der Providentia-Lehre
durch Bartholomäus Keckermann im Zusammenhang der
Emanzipation der Geographie aus der Theologie.
Ursachen und Folgen' (The new orientation of the
doctrine of providentia by Bartholomäus Keckermann
in connection with the emancipation of geography
from theology. Motives and results), *Z. Relig.
und Geistesgesch.*, vol 28, (1976), 123–32

Büttner, M. 'Beziehungen zwischen Theologie und Geo-
graphie bei Bartholomäus Keckermann. Seine
Sünden- und Providentialehre eine Folge der
Emanzipation der Geographie aus der Theologie?'
(Relations between theology and geography of
Bartholomäus Keckermann. Is his doctrine about
sin and providentia a consequence of the emanci-
pation of geographical thought from theology?)',
Neue Z. für syst. Theologie und Religionsphilos.,
vol 18, (1976), 209–24

2. SELECTIVE BIBLIOGRAPHY OF WORKS BY BARTHOLOMAUS KECKERMANN

1601 *Meditatio de insolito et stupendo illo terrae motu,
anno praeterito 8 Sept. 1601... (Meditation about
the strange and terrifying recent earthquake 8
September 1601...)*, Heidelbergae; another ed. 1602

1603 *Systema compendiosum totius mathematices, h.e.
geometriae, opticae, astronomiae et geographiae...
(Abridged system of complete mathematics, i.e. of
geometry, optics, astronomy and geography...)*,
Hanoviae; other ed. 1617; 1621; Oxonii 1660-1;
1661

1607 *Contemplatio gemina, prior ex generali physica de
loco, altera ex speciali de terrae motu potissimum
illo stupendo qui fuit anno 1601 mense septembri
(Two studies, one on the common physics of place,
the other on the special one of earthquakes,
chiefly the terrifying one that occurred in
September 1601)*, Hanoviae; another ed. 1611

1611 *Systema astronomiae compendiosum in Gymnasio
Dantico olim praelectum et duobus libris adornatum
(Abridged system of astronomy, lectures formerly
given at Danzig College)*, Hanoviae

1611 *Systema geographicum (System of geography)*,
Hanoviae; other ed. 1612; 1616, 194p.

1612 *Brevis commentatio nautica (Short nautical
treatise)*, Hanoviae

1613 *Systema systematum (System of systems)*, ed. by
J. Alsted, Hanoviae

1614 *Opera omnia quae exstant (Collected works)*,
2 vols, Genevae

1617 *Systema astronomiae libri duo (System of
astronomy)*, Hanoviae

*Dr Manfred Büttner, holder of three doctorates, Dr
rer. nat., Dr phil. and Dr theol., is lecturer in
geography at the Ruhr University in Bochum, West
Germany.*

CHRONOLOGICAL TABLE: BARTHOLOMAUS KECKERMANN

Dates	Life and career	Activities, travel, fieldwork	Publications	Contemporary events and publications
1572	Born at Danzig			Counter Reformation
1580-8	Attended schools at Danzig			
1589				Philipp Apian died
1588-92	Studied at Univeristy of Wittenberg			
1594				G. Mercator died
1595				Mercator's atlas
1592-1602	Studied at University of Heidelberg			
1601		Travelled to Switzerland	*Meditatio de... terrae motu (Meditation on earthquake)*	
1602	Professor of philosophy at Danzig College			
1603		Application of gen. Geography for schools Conceived his System of Geography First philosophico-theoretical foundation of Geography	*Systema compendiosum totius mathematices (Abridged system of the whole mathematics)*	
1609	Died as professor of philosophy at Danzig			
1611			*Systema geographicum (System of geography)*	

Emile Levasseur
1828-1911

JEAN-PIERRE NARDY

From 1870 to his death, Levasseur was one of the
leading international figures of science but of
his considerable publications only some historical
works are still known.   All his numerous contributions to geography are forgotten but he was a pioneer
figure in the modern revival of French geographical
studies.

1. EDUCATION, LIFE AND WORK

a. Career
Born on 8 December 1828 (Pierre) Emile Levasseur
was the son of a working jeweller in Paris.   After
a splendid primary education he went to the Collège
Bourbon in 1839, where Taine and Prévost-Paradol
were fellow pupils.   An intelligent and hard working
pupil, Levasseur achieved entry to the Ecole Normale
Supérieure on 30 October 1849, and there renewed
his acquaintance with Taine and Prévost-Paradol.
He was of a friendly disposition and his companions
included E. About and O. Gréard.   He found his true
vocation through the influence of Chéruel, the historian (and teacher of Fustel de Coulanges), and
turned from philosophical to historical studies.
Nevertheless he refused to specialize and became a
Bachelor of Physical Sciences in December 1849.   He
came second in the Ecole in 1852 and took up a teaching appointment at Alençon, where he at once began to
write two theses, one on the *Système de Law* and the
other, in Latin on the *Finances dans l'empire romain
au IVe siècle de notre ère*.   Having successfully
defended his two theses in June 1854 he at once became

a specialist in economic history.   He gained the
agrégation in 1854 and was appointed as a teacher of
rhetoric at Besançon and in 1856 in Paris.   In 1861
he succeeded V. Duruy, who became Minister of Education, as a teacher of history at the Lycée Napoleon.
He continued to be an active researcher and the
Académie des Sciences Morales et Politiques warmly
approved his works on the *Question de l'or (The
question of gold)* in 1858 and the *Histoire des
classes ouvrières avant la révolution (History of
the working classes before the revolution)*.   When
the Comité des Travaux historique (committee on
historical research) admitted him as a member, it
was primarily as an historian and economist that he
was welcomed.   His study of geography, especially
of its economic aspects, developed for methodological
reasons discussed on p83.   Obliged to teach at least
some geography he saw the neglected condition of the
subject only too clearly and in 1861 he sent repre-
sentations to V. Duruy on the need to strengthen
geography teaching, particularly in relation to the
life of his time.   As minister for education Duruy
reacted favourably and asked Levasseur to prepare
geography courses as part of the second and third
years teaching in his newly-organized Enseignement
Spécial (secondary industrial and commercial education).   Even more significant was the Ministry's
inspection of 1871 by J. Simon with L.A. Himly for
this resulted in a thorough reorganization of geography teaching.   In this development Levasseur was
influential.   From 1868 he gave a course on economics (Facts et doctrines économiques) at the Collège
de France, where in 1872 he was given a chair in

geography, history and political science, and finally
became administrator in 1903.   From 1871 he taught
geography and statistics at the Ecole Libre des
Sciences Politiques;   also in 1871 he joined the staff
of the Conservatoire des Arts et Métiers, where in 1876
he became professor of political economy and industrial
law.   From this time he became known for his learning
not only in France but internationally.

As a member of numerous historical and statistical
associations he helped to establish the International
Institute of Statistics in 1885.   As a geographer he
became honorary president of the Société de Géographie
and he was also a member of the societies in London,
Holland, Lisbon, Romania, Berne, Italy, Russia, Madrid
and Eastern Switzerland.   He was also associated with
various foreign scientific organizations, and became a
member of the Academei dei Lincei (Rome) in 1882, an
associate of the Royal Academy of Sweden from 1894, and
a corresponding member of the Academies of Sciences of
Prussia in 1900 and of Austria in 1902.   He attended
several congresses in Europe and went to Brazil in 1886
and his visits to the United States in 1876 and 1893
were followed by the publication of *L'Ouvrier Americain*
in 1897.   Through these varied travels Levasseur made
the acquaintance of most of the world's specialists in
the social sciences: he corresponded with many of
them, and exchanged statistical material with friends
in many countries.   When he died he was a scholar of
international fame, honoured by his own countrymen
as Grand Officier de la Légion d'Honneur two years
earlier, in 1909.

b. Personality

Levasseur's quite exceptional career covered sixty
years and all the time he retained his intellectual
vitality and an open-minded interest in all that he
saw.   He was not made famous by one dramatic discovery
of revolutionary scientific interest but was a devoted
scholar of wide reading with a helpfully pragmatic
approach.   He climbed Mont Blanc before publishing *Les
Alpes et les grandes ascensions* (1889) and used every
opportunity available to him as a member of the Société
d'Horticulture to study in minute detail the problems
of farm management.   His own experience was as vivid
as his concern with theory and gave him an exceptional
power of conceiving a broad synthesis, firmly grounded
on a mastery of history, geography, statistics and economics.   He had an apparently unlimited capacity for
work and for orderly arrangement of his material.   He
managed to follow several activities calmly, preparing
courses, reading articles and books, receiving in his
office of almost monastic austerity colleagues and
students with respect and interest and not infrequently
acquiring from them information of value which he recorded carefully.   This 'lay Benedictine' had not only
courtesy and concern for all the people he met but also
a high sense of scientific duty.   This was seen particularly in his refusal to accept any remuneration for
his preliminary work on the course of Panama Canal in
1879.   But perhaps his character was best summarized
in the *scrire et prodesse* of his teachers at the Ecole
Normale.

> Caractère bienveillant.   Moralité irreproch-
> able.   Gravité judicieuse.   Esprit curieux
> et droit.   Application énergique et régulière.
> Zèle infatigable.

2. SCIENTIFIC IDEAS AND GEOGRAPHICAL THOUGHT

Levasseur's great interest in geography came after a
thorough education in history and economics and was an
expression of his belief that all human sciences were
fundamentally interdependent.

A. SCIENTIFIC IDEAS

a. Determinism and liberty

Always fully aware of the scientific aspects of contemporary problems, Levasseur was a worker depending
on a rigorous analysis of authenticated facts, following the practice of Chéruel who said that 'every
assertion must rest on proof ... drawn from unequivocal
and sure texts' (Liesse 1914, 343).   Only on such a
foundation is it possible to put forward general laws
on the behaviour of societies.   Established as the
practice in history and economics, this method should
also be used in geography for 'if it aspires to be a
true science, it connects facts to general concepts and
to find these concepts it goes forward to the causes
of phenomena' (1872, 26).   Levasseur saw that determinism had its application to the social sciences but
he was careful to avoid the excesses of the positivists
who saw in the environment, race or circumstances of
the time the effective explanations of the qualities of
individuals and societies.   Having read Carl Ritter's
L'introduction à la géographie générale comparée, 1836,
he said that its author had a kind of 'fatalism' in
regarding varied forms of civilization as if they were
some kind of vegetation related to the local soil
(1872, 20).

While recognizing that any society experienced a
wide range of natural and moral influences, at the same
time any society was itself the maker of its wealth and
of its social order: the environment and the people
(or the societies) were at once subject to determining
influences and makers of their own circumstances.
Social sciences therefore had the responsibility for
unravelling the complex tissues of interrelations
between the environment, the people and their civilization as well as of discerning the behaviour of indivi-
uals themselves possessing liberty of thought and
action.   The interplay of determinism and liberty was
a problem solved by Levasseur in an original manner.
He did not favour, like some early nineteenth-century
French geographers, the probabilist ideas of Cournot
who regarded liberty as a kind of measurable chance,
nor did he agree with the great statistician Quételet
who attempted to find the 'average man', an abstract
conception derived from the mathematical evidence of
statistics.   He did not think that individuals or even
nations inevitably had 'a natural tendency to develop
in a certain way' but that they had freedom of choice
and that in using it they profited from their merits
and suffered for their faults (1891, 197-201).

Liberty was the main impulse in the evolution of
societies and therefore in social study it was difficult to use the experimental methods of the natural
sciences founded on immutable laws.   As such methods
had to be discarded, others must be sought.   This
enquiry began with an exhaustive study of the statistical sources available, and led to an assessment
of their social significance.   The method, which
he termed 'experimentally dogmatic', was in fact empirical as its validity depended on the gathering of

observations, in his work over-plentiful.   Nevertheless through his practice of such research methods
Levasseur became an expert on all the influences on
social matters, crucial or marginal, that he wished to
explain and his knowledge of the physical sciences, in
particular of geography, gave him insight on natural
causes while his knowledge of social sciences, with the
statistical data, gave him discernment on causes of
distinctly human origin.   But his adoption of such a
demanding method of research restricted Levasseur's
wish to provide controversial and committed views on a
solid factual foundation.

b. The defence of liberalism

Coming from an artisan family, Levasseur had direct
experience from childhood of economic and social problems and this was clearly reflected in his first two
publications on economic history, the doctorate theses.
In fact they were the first theses of their kind for
throughout the first half of the nineteenth century the
humanitarian ideas of the Revolution and the problems
imposed by industrial growth had induced the publication of many works showing socialist inspiration.   As
advocates of the planned redistribution of wealth, the
socialists were in conflict with such liberal economists as Turgot, Adam Smith, J.B. Say and Ricardo, who
favoured free enterprise and competition.   Confronted
with socialist ideas, the liberals were impelled to
give some scientific justification for their theories,
generally on a psychological and economic basis.
Levasseur in his two theses explained his own views
with marked clarity.   He rejected the non-experimental
'idealist' methods and ideas of the utopian socialists
and the theories of Marxism and made the defence of
liberalism his life work.   His primary concern was
with political economy, as an explanation of the
theories and laws of production, distribution, exchange
and consumption of resources.   As this was crucial in
all human life, all economic study must rest on geographical analysis of population, resources, production,
exchange and cognate features.   By the evolutionary
study of economic history it was possible to test some
of the economic laws used by geographers.

The scientific defence of liberalism ultimately
led Levasseur to undertake a vast study of all aspects
of society so that 'he thought that all sciences of man
living in a society are interdependent and anyone who
isolates them from one another is probably on the wrong
track' (A. de Foville, 1912, 36).   It therefore
follows that political economy, geography and history
are not three distinct sciences but three complementary
aspects of social science, still to be developed
and differing from the sociology of A. Comte, which
Levasseur regarded as ill-defined.   This new 'science
of society', founded on the trilogy of political
economy, history, geography, could only be valid if
these three subjects were themselves recognized as
sciences.   History and political economy acquired this
recognition through the application of an 'experimental
method' but so far in France there was no really
scientific geography with such a method, and this
Levasseur felt impelled to create.

B. GEOGRAPHICAL IDEAS

Searching for the causes of interdependence as a
geographer Levasseur deliberately abandoned the
descriptive geography of the seventeenth and eighteenth
centuries in favour of the approach of A. von Humboldt
and Carl Ritter.   The *Kosmos* of von Humboldt had
appeared in France in 1851 and in it he found, but not
to excess, the encyclopaedic character of a geography
whose scientific value lay in its contribution to the
understanding of great natural or human general laws.
Workers such as von Humboldt had given geography a
place in relation to other sciences.   But Levasseur
was influenced far less by the determinism of Ritter's
Géographie générale comparée, 1836.   He did not agree
that the varied natural environments could so influence individuals or human groups that they would
determine the varied forms of ways of life and civilization.   He was however convinced by the idea that
geography was in effect a study of the relations
between man, nature and levels of civilization.
Together these two great pioneers enabled Levasseur to
create a geography possessing a method of specific
analysis and thoroughly integrated in the general realm
of knowledge through its relation with other sciences.

Through analysis of the relations between man, the
environment and economic systems it was possible to
find geographical laws on the behaviour of man in
society.   Levasseur, arguing against determinism, said
that though the industrial wealth of a country depended
on its mineral resources the crucial fact was the use
people made of them.   People were studied as *homo
economicus*, for Levasseur deliberately chose to leave
aside their aesthetic, political or public spirited
motives and concentrated on their economic behaviour,
as this could change and evolve through human freedom.
As this freedom meant that economic behaviour was
never determined by any economic system, Levasseur
saw in geographical method a principle of 'evolution'
rooted in the liberty of humanity conceived as a kind
of mutative deviation in *homo economicus*.   Geography
could be regarded as 'experimental', for it was difficult to find geographical laws in a society permanently
in a state of evolution.   It was however possible to
choose one moment in time and analyse all the dated
statistical material (l'histoire au repos', 1889, in
'La population française', vol 1, 51) and look for the
abiding relationships retained through the changing
circumstances of each evolving society (history was
'statistique en mouvement' 1889, *ibid.*, 51).   If such
relationships existed in a variety of epochs and social
circumstances, then they could be regarded as geographical laws.

Levasseur's originality as a geographer rested on
the objective character of his statistical analysis and
his experimental method of studying historical evolution.   In his analysis of particular cases and in his
search for general laws, his method was at once ideo-
graphic and nomothetic.   It became something more than
a comparative geography, for through the 'experimental
method' it became possible to study both any single
geographical feature while also recognizing its evolution.   This acceptance of the evolutionary principle
showed that Levasseur had been well aware of the
scientific ferment of the nineteenth century though he
was not a supporter of any one of the rival theories
of the time.   For him evolution took place without
any clear and established plan, without any influence
of environment, possibly as the effect of some chance
circumstances though not by hazard, but by some form

of human liberty.

Of considerable interest is the methodology of
Levasseur, comparable but yet different from Ratzel's.
Levasseur saw the relation of geography to other
sciences, to mathematics, astronomy, physical and
natural sciences, statistics, history and political
economy.   In this view he was partly inspired by
A. Comte.   But though geography was at the summit
of this classification, it was not an encyclopaedic
synthesis of all these aspects of learning and
'though it acquired material from them it did so only
to express its own distinct point of view', which was
inevitably 'the study of the earth ... in all its
aspects, but always the earth' (1872, 49, 56).   This
gave the 'scientific point of view' to Levasseur's
approach: far from restrictive, it opened a vast range
of enquiry which he suggested should be followed by
specialists in different branches.   This would in the
end, provided that the main purpose was retained, preserve the unity of geography and indeed reinforce it.

Levasseur provided a geographical method suited
to study in both time and space, equipped to probe
phenomena of distribution, relationship and evolution.
It would not only study things past and present but
also, to some extent, look into the future with the
help of general laws.   But there was a need to be
cautious about prediction, as for example in studying
the Panama Canal for its course (so thoroughly discussed before its construction) must be a human choice
from so many apparent alternatives that dogmatism
would be foolish.

3. INFLUENCE AND SPREAD OF IDEAS

Through his interest in political economy and history
and in other social sciences, Levasseur was naturally
drawn to a study of the distribution of population,
particularly to a demographic geography.   Aware that
his method was new, he expressed it through teaching,
both by providing new courses and by experimental
methods of instruction.

a. Demographic geography

This was not new for Levasseur found much of interest
in the work of statisticians of earlier centuries.
From the sixteenth century there had been a conscious
relationship of geography and statistics, especially
well understood in the eighteenth century and notably
c. 1830, by the 'German descriptive school' inspired
by Achenwall.   There were similar studies in France
by Necker and Lavoisier.   From the beginning of the
nineteenth century systematic censuses had been studied
by statisticians such as Legoyt and especially Turquan,
to whose work Levasseur added a map of population of
a rural character for its time (1889 'La population
française').   The copious observations on population
from the eighteenth century by economists and other
scholars gave the challenge of discerning laws, and
Levasseur attempted this in a theory of population
based on the temporal and spatial variations related
to various natural and above all economic factors.
There was attraction and repulsion for populations in
differing environments, with the obvious avoidance of
harsh climates.   And in France, as E. de Beaumont had
observed, the Paris basin with its concentric topography had been more attractive to human settlement

than any other area while the broken topography with
no central area in the Massif Central had repelled
settlement.   In areas of dry soil people settled
where water rose in wells, springs and valleys, while
settlement was likely to be dispersed in humid areas.
Levasseur however knew that there were many exceptions
to these generalizations and believed that economically
developed societies were relatively independent of
environmental conditions.

He therefore proposed a series of 'laws of social
attraction', and in 1872 in his *Etude sur l'enseignement de la géographie* discussed the location of
industry;   heavy materials of small value were not
transported far because of the cost and so industries
using such resources were located near the source of
the materials or where they could be imported at ports
or beside rivers and railways.   And clearly many
industries, especially of commodities in frequent use,
would be located in densely populated areas where there
were the largest number of consumers and the lowest of
transport costs.   Levasseur was apparently the originator of this economic law later given precision by
A. Weber.   Also industries could be attracted by a
suitable labour force.   These various laws clearly
resulted in the growth of densely settled areas with
economic advance.   Levasseur formulated a law, reproduced in a less rigorous form by Max Sorre in the
Fondements de la géographie humaine (1951, vol 3,
294) 'towns ... have an attractive power on the rural
population which is in proportion to their population'
(1909, 56).   The theory of Levasseur also influenced
Vidal de la Blache, though he argued that populations
do not spread through a region like drops of oil but
rather that they concentrate in favoured areas, 'about
centres or along lines of attraction'.

In his geographical laws Levasseur was seeking
some relationship between population density and level
of civilization.   This he found in a theory of the
evolution of societies, originating among German historians and especially in the work of Roscher.   Five
stages could be seen.   The first, *sauvage*, was pre-
agricultural, with densities of 0.02 to 0.03 per square
km and the second *pastoral* with rudimentary agriculture,
1 to 2.7.   In the third stage, *agricole*, the agrarian
life brought densities of 10 to 40, with far higher
densities in the fourth stage, *industriel*, and finally
at the *industriel et commercial* stage even greater
densities would occur, depending on the intensification
of agriculture and of industry and the provision of
abundant transport.   With this progression from the
primitive to the advanced economic state of society,
direct dependence of natural environment was lessened.
Undoubtedly these theories showed the advance of geography in social studies and through this work Levasseur
showed the need for better education in geography.

b. Geography in education

Before the reforms instigated by Levasseur, geography
was sadly neglected in France.   In higher education
there was not a single chair;   in secondary education
there were only a few courses given by historians,
many of whom were incompetent and dependent on obsolete
books and other materials.   In some cases there was no
provision at all.   Only in the Ecoles Normales was a
serious effort made, despite the ignorance of the
teachers (1871 Rapport Général), but the effects were

sterile.   The main purpose was to memorize lists of places, altitudes and figures, all of which did little to develop intelligence and still less to provide anything useful in professional life later.   Levasseur's educational programme of 1864 gave him experience that he could use profitably in 1874, when the lamentable ignorance of geography from the High Command downwards was known to be one of the causes of French defeat in 1870.   Levasseur's investigations with L.A. Himly in 1871 were followed by the institution of a three year course based on France, Europe and the rest of the world.   In primary schools, instruction was to be given on the natural environments of France with a more general treatment of France as a whole, Europe and the world.   Then followed the non-European continents, and after that Europe in general terms.   The fourth year was used to give a general knowledge of France, and the fifth for detailed study of France, with a more thorough treatment of non-European continents in the sixth year.   Young people continuing education beyond the primary stage included a thorough study of France and its colonies.   This programme could only be carried out with suitable texts, and Levasseur himself wrote some of these, notably *La France et ses Colonies*. He also enlisted the help of local experts to write books for local study in the series *Les petites géographies des départements*.

Levasseur and Ch. Périgot, who was a teacher at the Lycée Saint-Louis and the Ecole de Commerce prepared the illustrations for these books, and the atlas associated with them, and the drawings were done by A. Vuillemin, who was part author of several atlases produced at this time.   Levasseur also edited several wall maps and globes, always insisting that they must bear the closest possible relation to features on the ground.   In the text books, the need was to provide illustrative maps.   Excellent as the aspirations were, they did not result in a complete revolution in the teaching of geography.   There was still a heavy emphasis on memory work and pupils learned by heart the location of *départements*, the course of rivers, lists of figures and other details.   The novel aspect was the search for the explanation of facts and of situations and this won the praise of Vidal de la Blache (1911, 457) who said that Levasseur 'à fait pénétrer plus d'air et d'intelligence', both in schools and among the general public.

Even so, the manuals that won Vidal de la Blache's approval now seem too analytical, if characteristic of education when they appeared.   Their general plan was 1) climates, 2) geology, 3) relief, 4) hydrography, followed by various aspects of geography, 5) historical, 6) political, 7) agricultural, 8) mineral, 9) industrial, 10) commercial, 11) administrative, 12) demographic.   This plan, apparently inspired by the scientific classification of A. Comte, supposedly connects each aspect with what has gone before but would be likely to result in a series of separate treatments lacking any general synthesis, even though each chapter might conclude with a synthetic summary. But the aspiration was always in mind and Levasseur obviously wanted to give all the facts and to discern their significant relationships.   While the analytical treatment gave the opportunity of mastering a mass of material the explanatory synthesis was less apparent.   The original aim proved difficult to

achieve but Levasseur came up against a problem seen later in works on regional geography, better able to discern the significant features than to master their synthetic characteristics.

By his advocacy of geographical study and the provision of texts, imitated and improved by others though their plan still remains to this day, Levasseur had a decisive influence on modern French geography, for it became attractive to a new generation of workers.   He had no specialist students himself but taught geography to those whose main interest lay in other subjects, though his writings on the subject were well used.   Vidal de la Blache absorbed his ideas on the density of population, though in general geographers were more interested in his historical than his demographic geography.   The demographic work, his most original contribution, was favoured by sociologists such as Comte, who from his 'law of town attraction' saw a sociometric indicator in the degree of organization of the urban network or Durkheim, who looked for a relationship between density of population and the degree of division of labour.   Geographers reading sociological works came across the demographic theories of Levasseur, in some cases without the realization of their original source.

Vidal de la Blache, in an obituary of 1911, said frankly that the scholarship of Levasseur had been superseded by scientific development though he praised his educational work.   Levasseur had written a eulogistic review of Vidal de la Blache's *Tableau de la géographie de France* in which he said (1903, 617), that during the previous fifty years geography had become a 'philosophical science'.   In the last analysis Levasseur, as an interdisciplinary scholar, had no direct influence for in the scientific world of his time he held a marginal position as an opponent both of strict determinism and of contingency, and not rigidly identified with any one science in which he made his mark.   But he was the founder of a scientific method of credibility and, by the authority he acquired as a scientist and by his formidable powers of advocacy, he achieved the universal acceptance of geography as a science in its own right, indispensable for the understanding of physical and human phenomena.   And he was a precursor of eminence, as well as of practical help, in the renewal of French geography.

Bibliography and Sources

1. OBITUARIES AND REFERENCES ON EMILE LEVASSEUR
Thieme, Hugo B. has compiled a bibliography with 216 titles of Levasseur which is included in *Bibliographie de la littérature française*, Paris (1933), tome 2, 154-8
Obituaries include:
Chuquet, A., 'Discours a l'occasion de la mort d'E. Levasseur', *Séances et Trav. de l'Acad. Sci. Morales et Polit.*, no 176 (1911), 125-32 and *Mem. Acad. Sci. Morales et Polit. Inst. Fr.*, vol 28 (1916), 563-72
Levy, R.G., 'E. Levasseur', *Rev. Deux - Mondes*, vol 5 (1911), 96-132

Leroy-Beaulieu, A., 'E. Levasseur', *Rev. Sci. Polit.*,
    vol 26 (1911), 845-57
Vidal de la Blache, 'E. Levasseur', *Ann. Géogr.*, vol 20
    (1911), 456-8; *Geogr. J.*, vol 38 (1911), 437-9;
    Bull. Soc. Geogr. Comm., Paris, vol 33 (1911),
    529-36; *La Géogr.*, vol 24 (1911), 273-6
de Foville, A., 'Notice sur E. Levasseur', *Séances et
    Trav. de l'Acad. Sci. Morales et Polit.*, no 177
    (1912), 27-45 and in *Mem. Acad. Sci. Morales et
    Polit. Inst. Fr.*, no 28 (1912), 459-80
Deschamps, A., 'E. Levasseur et l'économie politique
    en Conservatoire des Arts et Métiers', *J. Econ.*,
    no 40 (1914), 24-34
Liesse, A., 'Notice sur E. Levasseur', *Séances et Trav.
    de l'Acad. Sci. Morales et Polit.*, no 181 (1914),
    337-61
More recent studies are:
Fohlen, C. 'Levasseur' in *Int. Encycl. Soc. Sci.*, vol 9
    (1968), 261-2
Nardy, J.P., 'Levasseur Géographe', *Cah. Géogr.*,
    Besançon, no 16 (1968)
The report on geographic education which was so signi-
    ficant in Levasseur's work is Himly, L.A.,
    *Rapport général sur l'enseignement de l'histoire
    et de la géographie*, Paris 1871, 47p.

2. SELECTIVE AND THEMATIC BIBLIOGRAPHY OF WORKS BY EMILE LEVASSEUR

a. Theses and early works

1854 *De Pecuniis publicis. Quomodo apud Romanos quarto
    post Christum saeculo ordinarentur*, Paris, 85p.
1854 *Recherches historiques sur le système de law*,
    Paris, 408p.
1858 *La question de l'or, les mines de Californie et
    d'Australie* (thesis)
1859 *Histoire des classes ouvrières en France depuis la
    Conquête de Jules César jusqu'à la Révolution*,
    Paris, vol 1 587p., vol 2 560p.

b. The teaching of geography

1872 *Etude et l'enseignement de la géographie*,
    Paris, 126p. and in *Séances et Trav. de
    l'Acad. Sci. Morales et Polit.*, no 96 (1871),
    415-92
1874 'Les méthodes de l'enseignement géographique',
    Rev. Polit. et Litt., vol 6 768-74

c. Main textbooks and atlases

1866-75 *Cours de géographie a l'usage de l'enseignement
    secondaire spécial* (published under the direction
    of Levasseur), Paris
1866-75 *Cours complet à l'usage des Lycées et Collèges*
    (published under the direction of Levasseur),
    Paris
1868-73 *Cours complet de géographie. Enseignement
    secondaire* (published under the direction of
    Levasseur), Paris
1868 *Géographie de la France et de ses colonies*,
    Paris, 3 vol
1873-5 *Géographie des départements à l'usage de
    l'enseignement primaire* (published under the
    direction of Levasseur with the collaboration of
    local specialists), Paris
1874 *Géographie des écoles primaires*, Paris
1875 *Atlas universel de géographie* (with A. Vuillemin),
    Paris
1878-90 *La France et ses colonies*, Paris, vol 1 570p.,
    vol 2 690p., vol 3 512p.
1886 *Précis de la géographie physique, politique et
    économique de la Terre*, Paris
1886-7 *Cours de géographie. Enseignement secondaire
    des jeunes filles*, Paris
1889 *Les Alpes et les grandes ascensions*, Paris, 447p.
1889 *Le Brésil*, Paris, 100p. with vol of views
1900 *Atlas de géographie économique*, Paris

d. Economic geography and geography of population

1873 'Géographie économique', *Rev. Polit. et Litt.*,
    vol 4, 714-20
1876 'La population de la France', *Rev. Polit. et
    Litt.*, vol 10, 148-54
1887 'Les populations urbaines en France comparées à
    celles de l'étranger', *Séances et Trav. de
    l'Acad. Sci. Morales et Polit.*, no 127, 251-336
1888 'Statistique de la superficie et de la population
    des contrées de la terre', *Séances et Trav. de
    l'Acad. Sci. Morales et Polit.*, no 129, 93-110
1889 *La population française*, Paris, vol 1 468p.,
    vol 2 533p., vol 3 569p.
1890 *Grand Atlas de Géographie, physique et politique*,
    Paris
1901 'L'influence des voies de communication du XIXe
    siècle', *Séances et Trav. de l'Acad. Sci. Morales
    et Polit.*, no 155, 161-75
1909 'La répartition de la race humaine sur la globe
    terreste', *Bull. Inst. Int. Stat.*, vol 17 (2e
    livre), 48-63
1911 'Quelques conséquences du progrès des moyens de
    communication', *Rev. Sci. Polit.*, vol 26, 701-15
1911 'Les ports et la marine en France', *Rev. Econ.
    Int.*, vol 8 (3), 7-47

e. Articles on geography and other sciences

1868 'Du role de l'économie politique dans les sciences
    morales', *Rev. Polit. et Litt.*, vol 1, 121-8
1891 'Les lois de la démographie et la liberté
    humaine', *Séances et Trav. de l'Acad. Sci.
    Morales et Polit.*, no 136, 191-219
1901 'L'enseignement de l'économie politique au
    Conservatoire des Arts et métiers', *Rev. Int.
    Enseign.*, vol 41, 211-28, 294-305, 385-99
1903 'La géographie de la France par Vidal de la
    Blache', *J. Savants*, n.s., 1ère Ann., 617-26
1907 'Géographie et statistique', *Séances et Trav. de
    l'Acad. Sci. Morales et Polit.*, no 167, 716-20
1907 'Le Socialisme à l'oeuvre', *Rev. Polit. et
    Parlementaire*, vol 54, 153-60

*Jean-Pierre Nardy, agrégé in geography, is 'assistant'
at the Institut de Géographie of the University of
Besançon. Translated by T.W. Freeman.*

CHRONOLOGICAL TABLE: EMILE LEVASSEUR

Dates	Life and career	Activities, travel, fieldwork	Publications	Contemporary events and publications
1828	Born in Paris, 8 December			
1836				C. Ritter, *Géographie générale comparée*
1849	Entered Ecole Normale Supérieure	Fellow students included Taine, Gréard, Prévost - Paradol		
1851				Cournot, *Essai sur le fondement de nos connaissances*
1852	Teacher at Alençon			
1854	Teacher at Besançon		Doctorate theses (listed in text)	
1855	Teacher, Lycée St. Louis, Paris			A. von Humboldt's *Kosmos* published in French
1858		Award given by Acad. Sci. Morales et Polit.	*La question de l'or*	
1859			*Histoire des classes ouvrières*	
1861	Teacher, Lycée Napoleon, Paris	Appointed to committee of *Trav. Hist. et Soc. Sci.*		
1863		Worked on programme of *Enseign. Secondaire Spécial* Further award by Acad. Sci. Morales et Polit.		
1864		Third award from Acad.		
1866	Chevalier of Légion d'Honneur		Began publication of textbooks on geography	
1868	In charge of courses at Coll. Fr.	Member of Acad. Sci. Morales et Polit.		
1869				Quételet, *Essai physique sociale*
1871	In charge of courses at Conservatoire de Arts et Metiers and at Ec. Sci. Polit.	Tour of inspection with Louis-Auguste Himly		
1872	Professor at College de France			
1876	Professor at Ec. libre Sc. Polit. and at Coll. Arts et Métiers Visit to America	President, Comm. Stat. Enseign. primaire and Vice-President Comm. ... Stat.		

Dates	Life and career	Activities, travel, fieldwork	Publications	Contemporary events and publications
1878	Officier of Légion d'Honneur		La France et ses Colonies	
1883		President, Comité Trav. Hist.		
1885		Vice-President, Inst. Intern. Stat.		Congress in London
1887				Congress in Rome
1889			La Population Française	
1891				I. G. K. Congress in Berne
1892				Congress in Genoa
1893	Visit to America			Durkheim, *Essai de physique sociale*
1894				Congress in Budapest
1896	Commandeur of Légion d'Honneur			
1900			Last of the geography texts appeared	
1903	Administrateur, Coll. Fr.	Members, Cons. Supérieur Instruction publique		Vidal de la Blache, *Tableau de la géographie de la France*
1904		Vice-President, Comité Trav. Hist.		
1909	Grand Officier of Légion d'Honneur			Sociological Congress, London
1911	Died in Paris, 10 August			

Geoffrey Milne
1898–1942

KATHLEEN MILNE

Geoffrey Milne was the second of three sons of Sydney Arthur Milne, who was headmaster of a Church of England school in Hull. He was educated at Hymers College, Hull, and obtained scholarships to Leeds University where his interests apart from Science were in the Debating Society. He read and spoke French and German. His career at Leeds was interrupted by the First World War; he served with the Royal Engineers in France and was a member of the Carey Sandeman Force which was assembled to stem the German advance during the retreat of the Fifth Army in 1918.

He became interested in soils when he was appointed Assistant in Agricultural Chemistry to Professor Hendricks at Aberdeen in 1921 (from 1923 Assistant Lecturer), and in 1926 he returned to Leeds where he continued his interest under N.M. Comber, Professor of Agriculture. During his two years as Lecturer in Agricultural Chemistry at Leeds he added to the range of his teaching commitments, giving courses in soils, animal nutrition, and general agricultural chemistry. Both in Aberdeenshire and in Yorkshire he was an adviser to county agricultural staffs and he was particularly concerned with the Soil Survey of the Vale of York. He had his full share of degree and diploma examining and his interest in the training of agriculturalists was seen in his membership of the Agricultural Education Association from 1923 to 1942. He was an examiner for degrees and diplomas in agriculture. At Leeds he met Marian Kathleen Morgan, a former student of Professor H.J. Fleure at the University College of Wales, Aberystwyth who was a lecturer in the Department of Geography. They were married in 1930 and had two sons.

In 1928 Milne went to Amani, Tanganyika to develop soil research at the East African Agricultural Research Station. His work extended through Kenya, Uganda, Tanganyika, Zanzibar, Nyasaland (Malawi) and Northern Rhodesia (Zambia), in all of which he travelled widely. He also made short visits to Southern Rhodesia and South Africa, including the Universities of Witwatersrand and Pretoria and the Department of Agriculture at Pretoria and Onderstepoort. His knowledge of East African soils was greatly ahead of anyone else's at his time and has probably not been equalled since. His paper of 1935, 'Some suggested units for classification and mapping, particularly for East African soils', attracted wide interest and is regarded as a landmark in the development of soil classification for soil surveys. Recognizing that varieties of soils do not occur haphazardly, he proposed the establishment of new units for the mapping of soils -- the 'fasc' and the 'catena', the fasc (Latin *fascis* = a bundle) being defined as a group of soil series with similar conditions of pedogenesis but 'differing in their degree of maturity, in the effects of man's intervention in soil development, in the effects of changes of vegetation or in the sum total of small effects due to their being geographically distant in occurrence'. The 'catena' (Latin = a chain) is essentially a cartographic unit and consists of 'soils which, while they fall wide apart in a natural system of classification on account of fundamental genetic and morphological differences,

are yet linked in their occurrence by conditions
of topography, and are repeated in the same relationship to each other wherever the same conditions
are met...'. The term 'fasc' has not been adopted,
although a similar concept is in use in the term
soil 'family'. The catena, on the other hand, has
become a standard concept both in soil survey and
in the study of soil genesis, and there is a large
and growing scientific literature devoted to it.
     The results of his six years of study in East
Africa were published as the *Provisional soil map of
East Africa* (in collaboration with soil scientists
in Kenya, Uganda and Zanzibar) with an explanatory
memoir in 1936. The map was drawn during 1935 at
the Royal Geographical Society's House in Kensington
Gore, London. In constructing the map he used his
ideas about 'catenas' which were 'shown as units,
using stripes of two or more colours to show two or
more soil profiles occurring in a defined relationship to each other throughout the area so depicted'.
He was very conscious of the geographical aspect of
pedology and corresponded at length with Professor
H.J. Fleure on the subjects of soils and geography,
and also with Clement Gillman of Dar-es-Salaam who
was at heart a geographer. The 'Soil reconnaissance
journey through parts of Tanganyika territory' had
been written in 1936 as a Report and left in typescript. After his death in 1942 it was decided to
publish it as a memorial number of the *Journal of
Ecology*, edited by his wife and by Clement Gillman.
This is a masterpiece of environmental description
to which future generations of students may turn
with profit. In it he recognized that valley
floor (mbuga) soils store some of the fertility
transferred from the upper members of the catena;
his ideas on the utilization of these soils, on
'bringing fertility back uphill', may yet find a
place in agricultural development.
     In 1937 he was awarded, on the advice of the
Secretary of State for the Colonies, a grant from
the Carnegie Corporation Trustees to enable him to
visit the West Indies for four months and the
United States for three months for the purpose of
studying various aspects of tropical agriculture
and soil conservation. His report on the results
of these journeys was published in 1940. In the
West Indies he made acquaintance with agricultural
conditions in Trinidad, British Guiana, Tobago,
St. Vincent, Grenada, Antigua and Barbados. In
the United States he accompanied soil-survey or
soil-conservation parties in surveys in Indiana,
California, Arizona, New Mexico, Texas, the
Carolinas and Tennessee, and spent several weeks
working in the Soils Laboratories in Berkeley,
California.
     In 1940, he drew up a memorandum entitled
'Soil survey in East Africa: some desirable developments'. In this he stated that 'mistakes can as
easily be made in conservation practice, upon soils
that have not been thoroughly studied, as they can
and have been in agricultural or pastoral practice.
Soil surveys should precede the framing of conservation policy'. In January 1942 when he died, he
was visiting Nairobi with a view to taking up the
post of Scientific Secretary to the East African
Industries Technical Advisory Committee.

Bibliography and Sources

1. OBITUARIES AND REFERENCES ON GEOFFREY MILNE
Obituary notices appeared in *Nature*, Feb. 14, 1942;
     East African Agricultural Journal, April 1942;
     Tanganyika Notes and Records, June 1942.
Numerous references to his work appeared in the
     Report of the Conference on Tropical and Sub-
     Tropical Soils held at Rothamstead Experiment-
     al Station, Harpenden, England in June 1948.

*2. SELECTIVE BIBLIOGRAPHY OF WORKS
BY GEOFFREY MILNE*
1933 (*et al*) 'A chemical survey of the waters of
     Mount Meru, Tanganyika Territory', *Amani Mem.*
     38p.
1935 (*et al*) 'A provisional soil map of East Africa
     (Kenya, Uganda, Tanganyika and Zanzibar),
     Trans. 3rd Intern. Congr. Soil. Sci., vol 1,
     266-70
1935 (*et al*) 'A short geographical account of the
     soils of East Africa', *Trans. 3rd Intern. Congr.
     Soil Sci.*, vol 1, 270-4
1935 'Some suggested units of classification and
     mapping, particularly for East African soils',
     Soil Res. Br. Emp., vol 4, 183-98
1936 (*et al*) 'A provisional soil map of East Africa
     (Kenya, Uganda, Tanganyika and Zanzibar) with
     explanatory memoir', *Amani Mem.*, 34p.
1937 'Notes on soil conditions and two East African
     vegetation types', *J. Ecol.*, vol 25, 255-8
1937 'Essays in applied pedology. I. Soil type
     and soil management in relation to plantation
     agriculture in East Usambara', *East Afr. Agric.
     J.*, vol 3, 7-20
1938 'Essays in applied pedology. II. Some factors
     in soil mechanics', *East Afr. Agric. J.*, vol 3,
     350-61
1938 (*et al*) 'Essays in applied pedology. III.
     Bukoba: high and low fertility on a laterised
     soil', *East Afr. Agric. J.*, vol 4, 13-24
1940 'Soil and vegetation', *East Afr. Agric. J.*,
     vol 5, 294-8
1940 'Soil conservation -- the research side', *East
     Afr. Agric. J.*, vol 6, 26-31
1940 'A report on a journey to parts of the West
     Indies and the United States for the study of
     soils, February to August 1938', *East Afr. Agric.
     Res. Stn.*, Amani, 78p.
1942 'African village layout', *Tanganyika Notes and
     Records*, no 13
1943 'Some forms of East African settlements',
     Geography, vol 28, 1-4
1947 'A soil reconnaissance journey through parts of
     Tanganyika Territory, December 1935 to February
     1936', *J. Ecol.*, vol 35, 192-265

*Mrs Kathleen Milne was engaged in editorial and
information work at the Commonwealth Bureau of Soils,
Harpenden, until her retirement in 1967.*

CHRONOLOGICAL TABLE: GEOFFREY MILNE

Dates	Life and career	Activities, travel, fieldwork	Publications	Contemporary events and publications
1898	Born at Hull, 5 August			
1914				First World War
1917	Served in King's Own Yorkshire Light Infantry			
1918	Royal Engineers, Sound-ranging Section, France			End of war
1921	B.Sc. with first class Honours in Chemistry, Leeds University	University of Aberdeen, North of Scotland College of Agriculture, Assistant in Agricultural Chemistry		
1923	Member of the Agricultural Education Association	Aberdeen, Assistant Lecturer in Agricultural Chemistry		
1924	M.Sc., Leeds University			
1926	Member of Soil Survey Committee and of sugar beet Analysis Committee, Ministry of Agriculture (to 1928)	Lecturer in Agricultural Chemistry, Leeds University	Paper at Groningen meeting	International Society of Soil Science meeting at Groningen, Netherlands
1927		In acting charge of Agricultural Chemistry Department, Leeds University		
1928	Fellow of the Royal Institute of Chemistry	Examiner in Chemistry, Midland Agricultural College In charge of Soil Research, East African Agricultural Research Station, Amani		
1930		Acting Director of Amani station		
1932	Secretary of East Africa Soils Conference			First Conference on East African Soils
1934				Second Conference on East African Soils
1935			Paper at Oxford meeting *Soil Research*, paper on soil classification	International Society of Soil Science meeting at Oxford, England
1938	Carnegie travelling studentship to West Indies and United States			
1939		In charge of Coffee Research Station, Lyamungu, Mashi, Tanganyika (war emergency duty) to 1940	Assistant Editor, *East African Agricultural Journal*	Outbreak of Second World War

Dates	*Life and career*	*Activities, travel, fieldwork*	*Publications*	*Contemporary events and publications*
1941	Collaborated on soils with the Nutrition Unit, Nyasaland	Secretary of Vegetation Classification Committee, East Africa. Drew up memorandum for the vegetation map of East Africa		
1942	Died at Nairobi, January 16			

Wincenty Pol
1807–1872

JÓZEF BABICZ

Wincenty Pol was a poet who also held a chair of geography from 1849, the first chair of geography in Poland and the second in the world after that set up by Carl Ritter in Berlin in 1820.

1. EDUCATION, LIFE AND WORK

Pol was born at Lublin in 1807. He was a descendant of a German family whose name had been Pohl. His father, married to the daughter of a rich bourgeois family of French origin (Longschamps) but already Polonized, for a long time held various administrative posts in Galicia, during which time he became so attached to Polish tradition that it prevailed entirely in his home. The patriotic aspects of Pol's upbringing were further enhanced at the schools he attended -- the grammar school in Lvov and a Jesuit school of Tarnopol, where he studied philosophy -- and again at the University of Lvov. In 1827 he finished his studies with a good knowledge of Greek, Latin and world history. On the advice of a relation, Adam Tocher, a librarian in Vilna, he studied German literature, which enabled him in 1830 to become a deputy lecturer in German at Vilna University.

He participated in the Polish uprising of 1830 and as a result was forced to emigrate to Germany but on his return to Galicia in 1833 he was welcomed at the country houses of Polish aristrocrats at Zagórazny, Kalenica and Polanka. Under suspicion by the Austrian police he could not write as a political poet but was constrained to restrict himself to singing the beauty of his homeland. During his long stay in the Carpathians, therefore, he devoted much time to the study of their geography and also to geography itself, for he was moved by its neglect in Poland. This was at a time when Humboldt's ideas in geographical research were becoming famous and when the Polish pupils of Carl Ritter -- such as J. Kremer, the philosopher -- were spreading the views of their teacher. Wincenty Pol kept in close touch with Kremer and other Galician scientists, such as the botanists W. Besser, J.K. Lobarzewski, J. Warszewicz and A. Zawadzki, the ethnographer Zegota Pauli, the historian A. Bielowski, the meteorologist J. van Roy and the geologist L. Zejszner.

It was at this time that Pol conceived the idea of producing a modern geography of Polish lands based on field studies. Unfortunately his manuscript and map material was destroyed in 1846 during a peasant revolt against landlords. At this time of the 'Springtide of Nations' Pol's literary output was already considerable and he was famed as a poet, extolling Poland's past, but he had also written a few papers on geography and had acquired a vast body of knowledge of the subject. The Austrian government, influenced by the freedom movements, permitted Cracrow University to be repolonized and Pol, convinced that he could serve his country better as a geographer than as a poet, applied for the chair of geography. In this he exploited his acquaintance with the Austrian Minister of Education, Count L. Thun, and also won the formal competition for the post. In the favourable political circumstances of the time the Ministry of Education in Vienna agreed in 1849 to the establishment of a chair of 'physical and comparative geography'.

In his scientific work, especially in teaching, Pol adopted an intensely Polish patriotic attitude.

For this reason four years later, when the Austrian government reaffirmed its traditional anti-Polish policy, Pol was one of the first victims of the growing absolutism and germanization. Considered politically dangerous, he was dismissed with other Polish professors and left without any means of subsistence. His hopes and ambitions were completely dashed and for the rest of his days, ill and finally almost blind, he lived, with assistance from friends in the nobility, in Galicia, at Cracow and Lvov. Attempts to settle him on an estate bought for him by Polish society at Firlejowka, near Lublin, eventually failed because he did not obtain permission from the Russian authorities. During this period his geographical activity declined and his literary activity grew. Polish science recognized him by offering him membership of the Academy of Sciences shortly before his death at Cracow in 1872. Numerous publications on his life and work followed.

2. SCIENTIFIC IDEAS AND GEOGRAPHICAL THOUGHT

On taking up his professorial appointment Pol set about the work of producing a geography of Poland based on field investigations, an aim he pursued with much initiative and inventiveness. He classified landforms and the network of rivers and in the process developed a terminology of hydrology. Although he had at his disposal no statistical climatic data and no accurate hypsometric map, by relying on his knowledge of the earth's sculpture, climate and flora and drawing on information gained during his travels and his reading, he was able to define regional zones of Poland which have largely proved to be correct and similar to those in use at the present day. In plant geography he correctly outlined distributions in botanical areas, dependent on the earth's relief and climate. He paid especial attention to the collective life of plants -- associations -- which later became of particular interest to him.

In his detailed ethnographical studies of the Carpathian area he established the territorial range of ethnic groups and in doing so applied the so-called criterion of 'secondary evidence of cultural products', a method used in the second half of the nineteenth century by a representative of the theory of cultural circles, Friedrich Ratzel (1844-1904).

Pol apprehended the trend in European geography towards a scientific approach and himself endeavoured to remain strictly scientific in his work. But as a poet it was difficult for him to avoid using literary metaphors and in his descriptions of landscape he gave rein to these tendencies, for example in *Północny wschód Europy pod względem natury, III: Obrazy z życia natury (The Nature of Northeast Europe, part III: Pictures from Life and Nature)* (1869). In his study of the river network the measurable criterion of navigability was applied in his *Hydrography* (1952). In the more detailed text of *Rzut oka na północne stoki Karpat (A look at the northern slopes of the Carpathians)* he described in the then customary systematic way the relief, river network, climate, flora, fauna and, more comprehensively, the ethnic character of the population -- the last being a subject of special interest to him. Endeavouring to follow Humboldt he aspired to give an integrated account of all the elements of nature and human life, but the inadequacy of his descriptions of

the country's regions and of its division into natural regions indicated that he was a long way from achieving a complete scientific geography of Poland. Yet all his work was distinguished by an unusual perceptiveness, an insight into geographical phenomena -- a 'geographical sense' -- which led him into interesting fields of work, such as his study of the distribution of Polish dialects.

In the lectures he gave during the three years of his professorship, Wincenty Pol touched on a wide range of geographical subjects. Besides conducting courses on Polish geography, he lectured on the geography of the five continents, oceanography, the geography of the Austrian monarchy, of Syria and of Palestine, physical geography, ethnography and commercial geography. For these lectures he used foreign, mainly German, sources. He also gave lectures for teachers and talks for the general public, of which the latter were very successful because of his popularity and oratorical talent. In student teaching he went beyond lecturing by engaging them in practical exercises, observations and field excursions. Sometimes this involved direct scientific investigation: under his guidance students carried out hydrological observations of the Vistula, meteorological observations at Cracow, and barometrical measurements in the nearby Carpathians, including the measurement of the level of the astronomical observatory in Cracow. Other work in which he took a close interest was the delimitation of Slavonic ethnic groups, the drawing of an orographic map of the Tatra Mountains, and research on place names. That his interest in research was not purely academic was demonstrated when he initiated the establishment of a network of meteorological stations, and in his involvement with the Scientific Society of Cracow and his endeavours for the establishment of a Museum in Lvov.

As a professor of geography at Cracow University Pol was in a position to make more widely known his ideas on the nature of geography. These were much in accord with German concepts, particularly those of Humboldt and Ritter. He regarded geography as a philosophical science of high standing, based on material drawn from many natural and social sciences, but he also attached great importance to investigations in the field as a primary source of geographical knowledge. Later in his life he ascribed great practical significance to commercial geography when, after the failure of the Polish national risings of 1830, 1846 and 1863, the only fields in which Poles could work with relative freedom were those of economics and education.

In his historical and polemical works, such as *Historyczny obszar Polski (The historical area of Poland)* (1869), he defended his country's right to the lost lands. His arguments for this were derived from Ritter's concept of natural boundaries and from political writings by a Polish émigré F. Duchiński, as well as from Hegel's historiosophical theory which considered the development of civilization as a migration of human spirit from the east to the west, Asia being the cradle of humanity while the peak of mankind's development was achieved in the geographical environment of Europe. Though Pol was morally right in this polemic, the arguments he used were as weak and partial as those put forward by the geopoliticians

of the occupying powers, then and later.   An example
of his partiality was his tracing of the boundary of
eastern Europe not along the Ural mountains, as had
been done by Pallas, but along the rivers Dniepr and
Dvina, the confines of Roman civilization and the
Catholic religion.

3. INFLUENCE AND SPREAD OF IDEAS

Although the weakness of so many of his geographical
writings lay specifically in their literary character,
Pol's geographical ideas were widely accepted in Poland
simply because he was so popular and so beloved as a
poet.   He contributed to the development of Polish
science by raising geography to the level of a university discipline at a time when it was regarded as such
only in a few countries.   In his work and in his
three years of activity in a university he introduced
the most up-to-date ideas on geography held by European
scholars, and he inculcated in his students an
appreciation of the value of field research, an
activity in which he himself participated vigorously.
Simultaneously with these scholarly activities he was
participating in the struggle of the Polish people for
independence, pointing to the individuality of the
geography of their lands as justification of their
rights.   His initiatives and his ideas were subsequently taken up by the next generation of Polish
geographers, especially by Eugeniusz Romer (1871-1954),
who is generally regarded as the founder of the Polish
school of geography.

In spite of a century and a half of foreign rule
in Poland, the evolution of the discipline of geography
in that country kept pace with that in other European
countries.   In the achievement of this, Wincenty Pol
was a major figure among Polish scientists of the time.

*geografii na Uniwersytecie Jagiellońskim (Wincenty
Pol as professor of geography at the Jagiellonion
University),* Cracow, 1949.

Professor *Józef Babicz* is head of the department of the
History of Natural Sciences in the Institute of the
History of Science and Technology of the Polish Academy
of Sciences.

Bibliography and Sources

The published works of Wincenty Pol were issued under
the title *Dzieła Wincentego Pola wierszem i proza
(Wincenty Pol's works in verse and prose),* Lvov,
1875-8, vols 1-10;   the geographical writings are in
volumes 2, 4, 6 and 10.   His correspondence was published by K. Lewicki in *Wincentego Pola Pamietniki
(Memoirs of Wincenty Pol),* Cracow, 1960.   J. Babicz
edited 'Wyklady geografii miane w 1854' (Geographical
lectures delivered in 1845), *Studia i Materiały z
Dziejów Nauki Polskiej,* Ser. C., vol 6, (1963), 54-97,
as well as Pol's works *Etnografia północnych stoków
Karpat (Ethnography of the northern slopes of the
Carpathians),* Ethnographic Archieves, Polish Folk
Science Society, Wroclaw, 1966.

Pol's life and activites have been comprehensively
presented by M. Mann, in *Wincenty Pol. Studium
biograficzno-krytyczne (Wincenty Pol. A critical
biography),* Cracow, 1904-6, 2 vols. His geographical
work has been reviewed by S. Niemcówna, *Wincenty Pol
jako geograf (Wincenty Pol. His life and geographcal
work)* Cracow, 1923 (in Polish but with an English
summary), and by H. Barycz, *Wincenty Pol jako profesor*

Dates	Life and career	Activities, travel, fieldwork	Publications	Contemporary events
1807	Born at Lublin			
1815				Formation of the Scientific Society in Cracow
1820				First chair of geography set up in the University of Berlin by Carl Ritter
1827	Graduated at Lvov University			
1830	Deputy lecturer in German at Vilna	Took part in national Polish uprising; emigrated to Germany until 1833		
1833-46	Lived in Galicia	Collected material on geography of Poland		
1835		Fieldwork in Carpathians		
1845		Visited Moravia and Alpine countries on behalf of the Economics Society of Lvov	*Podróz do Tatr (Journey to the Tatra Mountains) Wykłady geografii (Geography lectures)*	
1846	Editor of the Ossolinski Library Publishing House, Lvov			Peasant revolt and destructions of Pol's MSS
1847		Visit to Germany, Bohemia and Austria on behalf of Ossolinski Library Publishers	*Muzeum natury we Lwowie (Natural Museum in Lvov); O Hucułach (On the Huculy); Listy z podróży naukowych po kraju (Letters from scientific travels)*	
1848				'The Springtide of Nations'
1849	Appointed to newly established chair of physical and comparative geography at University of Cracow			
1850			*Rzut oka na umiejetność geografii ze stanowiska uniwersyteckiego wykładu (Geography through the eyes of a university lecturer)*	

Dates	Life and career	Activities, travel, fieldwork	Publications	Contemporary events
1851			*Opis Dniestru (Description of the Dniestr River); Rzut oka na północne stoki Karpat (The Northern Slopes of the Carpathians); Północny wschód Europy pod wzgledem (The nature of North-east Europe); Hydrografia (Hydrography)*	
1852			*Zasługi Długosza pod wzgledem geografii (Długosz as a geographer)*	
1853	Dismissed from chair of geography			Persecution of minorities in Poland by Austrian government
1856	Attempted to settle at Firelejowka, Lublin, unsuccessfully because of hostility of Russian authorities			
1863			*Geografia Ziemi Swietej (Geography of the Holy Land)*	
1866			*Dwie prełekcje o potrzebie geografii handłowej (Two talks on commercial geography)*	
1869			*O diałektach mowy połskiej (On the dialects of the Polish language); Historyczny obszar Połski (The Historical Area of Poland)*	
1872	Died at Cracow	Member of the Academy of Sciences, Cracow		

Carl Ortwin Sauer

1889–1975

JOHN LEIGHLY

In the heading of its report of Carl Sauer's death on 18 July 1975, the New York *Times* called him 'Dean of Geographers'. Whatever in Sauer's long life might have justified that title, it would have brought only a quiet smile to his face, as did the numerous distinctions awarded him during his lifetime. Attentive to the unanswered questions he always saw before him -- 'Life is short and there is terribly little time to learn the things that you want most to learn' (1938) -- he accepted recognition, as he wrote in acknowledging the award of the Victoria Medal of the Royal Geographical Society a few months before his death, only as 'an honest worker in the vineyard; perhaps ... one undisturbed by passing fancies'.

1. EDUCATION, LIFE AND WORK

Sauer's ancestors were members of a German pietistic sect that under the designation 'German Methodists' became a part of the general Methodist organization in the United States. His grandfather was a minister of the sect, his father a teacher in Central Wesleyan College, the German Methodists' institution of higher learning, situated at Warrenton, Missouri, where Sauer was born on 24 December 1889. His family heritage was one of simple piety, love of music, respect for humane learning, and, by the standards of the American Middle West, of permanent residence on native soil. His heritage from the wider community into which he was born was of 'the native values of rural life': industry, thrift, farsighted care for the land and for movable possessions. Sauer celebrated these values in his address, 'Homestead and community on the Middle

Border', 1962, one of the most appealing of his writings. Except for formal religious observance, they were the values by which he lived, taught, and wrote. His wife, Lorena Schowengerdt, to whom he was married on 29 December 1913, and who died only a month before his own death, was of the same small town and of similar ancestry.

Family connections with the land of their ancestors were sufficiently close to impel Sauer's parents to place him, in his ninth year, in a school in Calw, a small town in Württemberg. Five years of rigorous instruction there, with special emphasis on the classical languages, gave him a better fundamental education than he could have obtained at home. It enabled him, after his return to Warrenton, to complete the work for his bachelor's degree at Central Wesleyan College in 1908. He read widely in the college library, finding his greatest intellectual stimulation in the publications of the United States Geological Survey and of the comparable state surveys. His interest in geology took him first to Northwestern University, Evanston, Illinois, for graduate study of that science. There he found that petrography was emphasized rather than those aspects of geology which would interpret the earth as he saw it round about him; and when he learned of the instruction in geography being given by Rollin D. Salisbury (1858-1922) at the University of Chicago he transferred from Northwestern to Chicago in 1909.

If one looks in Sauer's works for influences of what he was asked to learn as a graduate student at Chicago, the most persistent ones to be found are insistence on the primacy of observation in the field and close attention to the forms of the earth's surface and to vegetation. These reflect the teachings of Salisbury, who headed the department of geography but whose courses were listed in the university catalogue under geology, and of Henry C. Cowles (1869-1939), a pioneer of plant ecology in the United States, whose field excursions led students to interpret vegetation as Salisbury's led them to interpret the forms of the lands. Sauer's apprenticeship in independent field-work was served in the summer of 1910 in the upper valley of the Illinois River. Nominally he was under Salisbury's supervision; but Salisbury gave him no instruction concerning what he should look for or how he should interpret what he saw. This insistence on independent work by students is one quality of Salisbury's teaching that Sauer retained in his later dealings with his own students.

Long afterward Sauer wrote to a correspondent: 'Most of the things I was taught ... as a geographer I had either to forget or unlearn at the cost of con-siderable effort and time'. His interpolation in this statement of the phrase 'as a geographer' is to be noted: Salisbury was of the department of geology and Cowles of botany. Without attempting to distribute Sauer's implied reproach among his instructors in geography, one may safely assume that a part of it was intended for the 'anthropogeography' of Ellen Churchill Semple (1863-1932), who lectured at Chicago during Sauer's years as a student there. What was taught as human geography in Sauer's days at Chicago was a rather simple mechanical theory of behaviour in which human beings 'responded' in various but primarily economic ways to the qualities of their physical surroundings. Only later did he sense a need to construct a more sat-isfying frame of reference for the observed facts of human geography.

In accordance with custom at Chicago, Sauer wrote his doctoral dissertation on a 'region', the Ozark Highland of Missouri to the south across the Missouri River from his native district. He worked for a year in Chicago as map editor for the Rand McNally firm of map publishers, and taught for a year at the then State Normal School at Salem, Massachusetts. On receiving his doctoral degree in 1915 he became instructor in geography at the University of Michigan, Ann Arbor, where he remained for seven years, attaining the professorship in 1922.

In Michigan he became acquainted with the formerly pine-covered lands of the northern part of the Lower Peninsula of that state, which after being stripped of their timber and often burnt over were of little use for agriculture. A large fraction of them had re-verted to the state for non-payment of taxes, and so had become an administrative 'problem'. Sauer and others concerned with the problem persuaded the state to undertake a 'Land Economic Survey' of the qualities and potential utility of the cut-over lands, which began work in 1922 by the experimental mapping of one county. In 1923 Sauer accepted appointment as professor of geography at the University of California, Berkeley, and so ended his connection with this Survey.

At Berkeley, in a part of the country wholly unfamiliar to him, Sauer took charge of a tiny depart-ment, staffed only by himself and two 'associates', instructors not yet qualified by the doctorate for regular appointment. The department had previously been little more than an appendage of the much larger and nationally prominent department of geology. Immediately on arrival Sauer had to appeal to the university administration as the geologists had spread into -- indeed encroached on -- the limited area available for geography before his arrival. He had also to reorganize instruction in the department to provide a broader range of courses than had previously been offered. He initiated a new sequence of intro-ductory courses, the first in physical geography built about the world distribution of climatic regions and the second organized by cultural regions. Finding in German geography the best example to be imitated, he brought to the department, at different times, Oskar Schmieder and Gottfried Pfeifer, and as visitors Albrecht Penck and Wolfgang Panzer. His department always remained small in comparison with most depart-ments of geography in American universities; he aimed at quality, not quantity, in both instructors and students.

Sauer had also to find new directions for his original work. Before leaving Michigan he had become dissatisfied with the formulation of the task of geo-graphy he had learned at Chicago, the rather abstract definition of the relations that subsist between human groups and their environments, the environment being conceived as the independent variable. He sought a more concrete formulation, one that would emphasize the facts acquired by observation rather than abstract 'relations'. Soon after he arrived in Berkeley he attempted to devise a program directed toward that end. The result of his efforts was *The morphology of landscape*, a scheme for the organization and inter-pretation of geographical facts. These facts can be divided into two classes, natural and cultural, both of which, but in association, undergo change through time. Sauer's emphasis was still 'regional', in accordance with the Chicago tradition, and the struc-ture he proposed resembled the one Otto Schlüter had offered twenty to twenty-five years earlier, though at the time Sauer had not read Schlüter's writings.

The morphology of landscape attracted much atten-tion among academic geographers in the United States, but exerted little influence on the substantive work Sauer and his students did after its publication. By 1925, when the essay appeared, Sauer had discovered Baja California as a rather easily accessible region where little scientific work had been done, and where a dry winter season made poor roads passable during the long winter vacations the University of California then enjoyed. For many years Sauer and his students used these vacations, and more time during winter when possible, for fieldwork in Baja California, northern mainland Mexico, and adjoining parts of Arizona and New Mexico.

In Mexico Sauer found a great variety of things that aroused his curiosity, especially in the more re-mote parts of the country; they led him into inquiries such as he had not entertained before, geographically farther southward, and chronologically into the past. He later described his reaction to Mexico in these words: 'I started in on Mexico, thinking I would go back no further than Spanish colonial institutions; and then I found myself back at the origin of man and his cultures'. The regional aspect of geography had little pertinence to his new investigations: he attempted no regional monograph after *Geography of the Pennyroyal* (1927), written up in Berkeley from material he had brought west with him, and discouraged his students from attempting them. Henceforth, fieldwork was to be directed toward the solution of problems, not toward the characterization of 'regions'.

The list of Sauer's published writings exhibits the range of questions his fieldwork in Spanish America raised in his mind, questions concerning historic and prehistoric settlement indicated by archaeological remains, and movement and occupation in Spanish time but ultimately and most persistently native agriculture and the plants cultivated. His observations of native cultivation and the keeping of the young of wild animals as pets prompted him to extrapolate backward to 'the origin of man and his cultures'. In several papers published in the late 1930s and 1940s he pursued inquiries into these origins. His studies culminated in *Agricultural origins and dispersals* (1952), a bold and original reconstruction of the remote past of humanity. Later, in 1970, he summarized his ideas concerning early man in the long essay 'Plants, animals and man'.

In 1954 Sauer relinquished the chairmanship of his department, which he had retained much longer than is customary at Berkeley, as he was reluctant to yield it into less competent hands than his own. He had already begun work on his next enterprise, the organization of the Symposium on Man's Role in Changing the Face of the Earth held at Princeton, New Jersey, in June, 1955. In the autumn of 1953 William L. Thomas, Jr., then assistant to the director of the Wenner-Gren Foundation for Anthropological Research, invited Sauer on behalf of the Foundation to act as chairman of such a conference, to be held in the summer of 1954. Sauer gladly accepted the invitation, but proposed that the conference be postponed for one year to gain time for thorough preparation. Sauer's fertile mind and Thomas's organizational talent enabled them to bring together a remarkable international group of scholars; the Symposium was a high point in the careers of all its participants. Its proceedings, edited by Thomas under the title of the Symposium, is an impressive volume of lasting value.

During the last twenty-odd years of his life the sources of Sauer's writings were from his reading and from the intellectual harvest he had accumulated during 'those brief and precious younger years when he (was) physically able to follow his clues' in the field. The first and best of these essays is *The early Spanish Main*, an account of the Caribbean lands at the time of Columbus's discovery and the consequences of the rapacity of the Discoverer and his successors. He next undertook, in *Northern mists*, to put together the scanty evidence of European knowledge of North America in pre-Columbian time. Then, in *Sixteenth century North America* and *Seventeenth century North America*, the latter not yet published at his death, he collected first-hand reports of what Europeans found of nature and native culture in the parts of North America they visited. These volumes were intended to define, as well as contemporary writings permit, the condition of the continent that Europeans were to occupy, the landscape on which they were to inscribe their cultural markings. At his death he left a fragmentary outline of his next undertaking, an historical geography of the

United States to have the title 'Recessional at the Bicentennial'. His notes show that in this work he would have laid particular emphasis on the prodigality with which his compatriots have squandered the native riches of their land, the excessive use and waste that now confront them with what may be the most critical problems yet encountered in their history.

2. SCIENTIFIC IDEAS AND GEOGRAPHIC THOUGHT

Although most of Sauer's writings are concerned with historical geography, his interpretations owe most to habits of thought he acquired through Salisbury's teachings in geomorphology and H.C. Cowles's in plant ecology. Few traces can be found in them of the ideas in economic geography and 'anthropogeography' to which he was exposed as a student. When he formulated the ideas he set forth in *The morphology of landscape* he extended to the visible marks of culture on the face of the earth methods of observation and classification traditionally applied to the forms produced by natural processes. Then, when he discovered the wealth and variety of material that Mexico and other parts of Hispanic America offered to his observant and curious eyes, he left behind him the hampering limits he had drawn in the *Morphology*. He adopted the simple procedures long used in studies of nature, applying them, *mutatis mutandis*, to the visible products of culture. He defined his unhampered view of the task of geography in a letter to Wellington Jones in 1934: 'I do think that our scientific end is an understanding of how things came to be. Physical geography does seem to be natural history and human geography culture history'.

Publication of *The morphology of landscape* led to a request for Sauer to write a few further statements about geography, which he used to wipe out, if possible, any lingering traces of the environmental determinism that had dominated much of geographical writing in the United States since 1900. Once that task was done, he had no further interest in methodology, declining even to discuss it in his correspondence. 'Method' is whatever needs to be done to achieve understanding of how anything about which curiosity has been aroused 'came to be'. That formulation implies a time dimension and processes acting along the axis of time. 'To me', he wrote to a correspondent in 1948, 'successions of events are the only feasible means of understanding the processes of geographic differentiation, be these differences physical or cultural. Changes in land forms, in climates, in vegetation are (processes of) natural history. What has happened to man around the world is intelligible to me only in terms of historical experience ... Origin, change, extinction can be illuminated only by joining when and where'.

The object of investigation he saw offered to himself and his students was, in the words of another letter, 'a physically differentiated earth, in itself undergoing change, subjected to differing occupations by culturally differing populations'. 'No man can encompass all this', he conceded; but it offers opportunities for a great variety of talent and special competence. His own interests during the 1940s and 1950s, after he had seen much of Hispanic America, were directed mainly toward the relations that obtain between human societies and plants. He

concerned himself with both constructive relations --
cultivation, selection, cultural dispersions -- and
destructive ones: the changes wrought by human beings
on natural vegetation, especially by use of fire.
The theme of human destruction of woody vegetation by
fire and the consequent increase in the area covered
by grass recurs in his writings through many years.

The great variety of questions raised by Sauer's
inclusive view of the physical and cultural differentiation of the earth's surface is reflected in the
topics on which his students wrote their doctoral
dissertations. He always encouraged his students
to choose the objects of their investigations according to their own interests and competences, exercising
his authority mainly to dissuade them from writing
general descriptions of 'regions', the kind of dissertation usually offered in American departments of
geography during the 1920s and 1930s. He welcomed
students whose undergraduate studies had been pursued
in fields other than geography, for they were likely
to have ability that enabled them to carry out
original, unconventional researches. One of the best
dissertations written under his direction was a masterly review of the fresh-water fishing practices of the
North American Indians. Its author, who had a degree
in anthropology, called it a study in 'historical
economic geography'. One student, who had a degree
in chemistry, investigated the indigenous alcoholic
beverages of Mexico; another, who was competent in
handling small sailboats, studied -- and sailed in --
the native watercraft of western South America.
Sauer encouraged his students to acquire useful
special abilities elsewhere in the university, never
entertaining the illusion that one small department
could impart all the knowledge a student needs.

The variety of interests Sauer pursued inevitably
brought him into scholarly relations with colleagues
in disciplines other than geography. In his early
days he thought that the closest intellectual relations of geography were with the 'social sciences',
especially economics. But he soon abandoned that
view; the great variety of cultures -- including
economies -- on the earth precludes universally valid
economic or social generalizations. He rejected most
emphatically the economists' notion that the techno-
logical-industrial culture of Europe and North America
is appropriate to all the peoples on earth. The
spread of the North Atlantic economy to 'undeveloped'
countries, as he saw it in Latin America, was radically
destructive to both the resources and the peoples
affected. 'Development' could have no other consequences for the inhabitants of the 'underdeveloped'
countries than to drag their inhabitants into the pro-
spective débâcle that awaits industrial culture when
the mineral resources it consumes so prodigally are
depleted.

On the Berkeley campus Sauer established
especially close collaboration with the ethnologist
Alfred Kroeber and the historian Herbert E. Bolton,
of whom he wrote in 1932, 'Probably there are few
geographers with whom I would have so much in common'.
With Kroeber and Bolton he founded the serial *Ibero-
Americana* as a medium for publishing works on the
culture history of Latin America, in which Sauer and
his students published much of their work. His
interest in cultivated plants led him into association

with botanists, not only those at Berkeley, but also
and especially with Edgar Anderson, of the Missouri
Botanical Garden, St. Louis. In his later years
Sauer thus established cordial relations with natural
scientists, including geologists and paleontologists,
among whom he found an appreciation of change through
time that articulated with his own thinking. The re-
semblance between these collegial associations and his
relations with Salisbury and Cowles in his student days
will not escape the reader. One of his last services
to his department, in the late 1950s when the university was building new quarters for various departments,
was to arrange that the department of geography be
placed in the new Earth Sciences Building, in comfortable proximity to geologists and paleontologists.

Although Sauer, after his brief venture into
methodology in the 1920s, avoided further methodological pronouncements, his ceremonial addresses before
the Association of American Geographers, 'Foreword to
historical geography' (1940, published 1941) and 'The
education of a geographer' (1956), passages in his
other writings, and his correspondence expound in
sufficient detail his matured view of his work and the
objects toward which it was directed. In 1951, in
reply to an inquiry by a student, he defined his
interests thus: (1) earth history in human times:
physical changes in the environment, such as climatic
alterations and changes in sea level; (2) man as an
agent in modification of the surface and vegetation
of the earth; and (3) migrations and blendings of
cultures and the forming of new culture patterns,
studied in terms of their geographic dynamics: direction, extension, and time. Though most of his work
was concerned with human geography, he encouraged the
cultivation of physical geography by his colleagues and
students; physical geography was emphasized in his
department more than in most departments of geography in
the United States.

For Sauer the basis of all sound work in geography
was observation in the field; he had nothing but scorn
for 'geographers who work in their offices through the
years when their legs, hearts, and eyes are good'. He
was himself a superior observer, blessed with the first
prerequisite of field observation, excellent eyesight.
'To me', he wrote to a former student in 1954, 'the
complete geographer is an eager and qualified field
observer, who can have his eyes communicate with his
brain so that relevant questions arise'. And to
another correspondent: 'I still think I can learn
more by being in the field than by reading. When I am
fresh from the field I have a new incentive to read,
and after I have read for a time I have additional
reason for getting back into the field'. He judged
students mainly by their ability to see things that
were before their eyes, and lamented that so few
acquired that ability. He looked on the field excursion as the best method of teaching, and took students
with him when making his own investigations, both in
order to inculcate habits of observation and to help
them toward independent work. 'We need people who
teach geography by doing it, and who will draw people
to work with them because they are enthusiastically
working themselves'.

Sauer once characterized himself as 'an earth
scientist with a slant toward biogeography, of which
man is a part'. That brief expression, however, says

little about the connotations his work had for him.
Life -- 'of which man is a part' -- is unthinkable
without the inorganic earth; but the surface of the
inorganic earth is far different from what it would
be without the living things that survive and per-
petuate themselves in the multifarious ecological
niches its surface firm and fluid envelopes provide.
In accordance with the teaching in plant ecology he
received as a student, Sauer saw this complex 'bio-
sphere' as always tending toward homeostatic balance.
But he came to doubt the simple concept of 'climax'
he had been taught, since both the inorganic and the
organic constituents of the ecological system are constantly changing. Here, as in all phenomena, time is
unidirectional; the climax appears to be static only
in the short view. Among living beings, man acquired
in time the means of changing his habitat more dras-
tically than any other organism can. The first
effective tool for changing the landscape man learned
to use was fire, to which Sauer attributed changes in
vegetation of a magnitude he thought were not yet
fully recognized. Later inventions enabled man to
modify the surface of the solid earth in permanent
ways: implements of tillage, mining technology, earth-
and rock-moving machines that operate on the scale of
such natural agents as earthquakes, floods, and
volcanic eruptions.

Patient and forgiving nature heals the wounds
inflicted by earthquake and lava flow, and eventually
may heal those inflicted by the earth's 'ecologic
dominant', as Sauer called the human species. But in
the meantime man may make the perpetuation of his own
species on the earth impossible. Since man is conscious of his actions, and can to a degree foresee
their consequences, he has a moral responsibility to
his posterity and his fellow creatures that cannot be
imputed to the earthquake and the volcano. The im-
perative posed by that responsibility, to preserve the
integrity of the ecological web of life, was the final
conclusion Sauer distilled from his contemplation of
man's tenure of the earth, a conclusion made urgent by
the reckless pillage of the earth's resources he saw
accelerate during his lifetime. In the end, the
essence of his teaching was moral. The wastefulness
of our culture has serious economic consequences, but
these are not the premises from which man's tenure
of the earth is to be judged. 'The moralist', he
wrote in concluding his address 'The education of a
geographer', 'lives apart from the quotations of
the market place and his thoughts are of other values'.
And elsewhere: 'What we need ... is an ethic and
aesthetic under which man, practising the qualities
of prudence and moderation, may indeed pass on to
posterity a good Earth'.

3. INFLUENCE AND SPREAD OF IDEAS
It was a source of embarrassment to Sauer that his
American colleagues in geography paid more attention
to *The morphology of landscape* and his other program-
matic writings from the middle 1920s than to his
substantive contributions. Those few methodological
utterances, though distinctive, represented only one
among several efforts to define for an obstinately
unspecialized and diffuse field of interest a neatly-
bounded place in a specialized academic world.

Although Sauer had no further interest in them, those
writings worked as a ferment in American academic
geography. Their most conspicuous effect was to
encourage detailed descriptions of small areas that
could be carved out of the map and identified as
distinctive by some physical or economic criterion;
descriptions of all aspects of such areas that appeared 'significant', usually significant with respect to
the current economy. Sauer saw both natural and
cultural landscapes as entities that have arrived at
their present state by changing through time; but by
others his emphasis on observation of details was
merged with an interpretation of the natural landscape
as merely the ground of economic activity, and of
cultural artifacts as elements in the contemporary
economy.

Sauer reacted strongly to this distortion and
impoverishment of his views, rejecting an essentially
descriptive frame for his own, and his students' work.
Though he published no more methodological writings,
his correspondence from the 1930s contains many denun-
ciations of a 'regionalism that is descriptive without
being analytic', which 'lacks criteria for determining
the diagnostic elements as to a genetic or dynamic
study'. 'Descriptive work must have an intellectual
objective'. Regional geography can be written, but
it 'has meaning only as a study of culture areas, and
the culture area is labelled and evaluated in terms
of what its occupants have done with the land they
possess'. The essential problem of regional geography
is 'the understanding of the mode of living and looking
at life of a particular group of people'. Gaining
such an understanding is immensely difficult, and
requires a long time: 'I think that the passing of a
cultural boundary is a terrific task for a geographer,
who must somehow learn to look through the eyes of the
people ... he is studying as well as his own'. '(A)
lifetime is just long enough to really get into the
problems of one region'. Whether these strictures had
any effect elsewhere than at Berkeley is difficult to
judge. In the United States the writing of detailed
descriptions of small areas went out of fashion at
about the time of the Second World War; but the new
fashion that followed the war owed nothing to Sauer.

Most of the things Sauer found interesting in
Latin America and on which he wrote were outside the
range of interest of the majority of his colleagues in
American academic geography, but his discussions of
them attracted the favourable attention of scholars in
other disciplines. Historians of Mexico and of the
North American Southwest in general appreciated his
combination of observation on the ground with the information supplied by Spanish documents. The editor
of the *Handbook of South American Indians* (7 vols,
1946-1959), Julian Steward, who as a graduate student
at Berkeley had known Sauer, invited him to write two
chapters for it, 'Geography of South America' and
the more important 'Cultivated plants of South and
Central America', the latter a long step on the way to
Agricultural origins and dispersals. Sauer's forays
into prehistory, both of North America and in general,
did not find so ready an acceptance among the specialists in a branch of learning in which evidence is
scanty. His repeated insistence that human beings
have been in North America for a much longer time than
had previously been supposed has been vindicated, and

some evidence has been found in southeastern Asia that supports his inference that that part of the earth was a primordial centre of domestication of plants and animals.

Sauer's self-esteem did not require the approbation or support of his colleagues in any discipline. His intellectual distance from his American colleagues in geography troubled him not at all. 'Formal bonds of membership in a particular academic fraternity', he wrote in a draft of an address in 1938, 'grow weaker as intellectual returns supply a sounder satisfaction of the desire for prestige'. He corresponded freely and cordially with his American colleagues in geography, but most of them conceived their métier more narrowly than he did, and his widely-ranging mind found few points of contact with theirs.

Though Sauer's work had little in common with that of most American academic geographers, his reputation attracted many students from other institutions to Berkeley for advanced study. More than half of the thirty-five students who wrote their doctoral dissertations under his guidance had received their first degrees at other universities. The oft-encountered designation 'Berkeley school of geography' was not invented at Berkeley, but came into use elsewhere to designate the source of a historical quality in writing and teaching, in contrast with the more usual emphasis on contemporary regional economies. Berkeley graduates readily found positions in other North American universities, and perpetuated there the kind of geography they learned from Sauer. Though 'the contemporary scene' dominates academic geography in the United States, there is a small but fairly steady output of studies in historical geography, by no means all by former Berkeley students, that would probably not be written without the example set by Sauer and by both his immediate and indirect disciples.

Bibliography and Sources

1. OBITUARIES AND REFERENCES ON CARL ORTWIN SAUER

Parsons, J.J. 'Carl Ortwin Sauer, 1889-1975', *Geogr. Rev.*, vol 66 (1976), 83-9

The following include bibliographies:

Kramer, F.L. 'Carl Ortwin Sauer, Geographer (1889-1975)' *Geopub. (Rev. Geogr. Lit.)*, Tualatin, Oregon, vol 1 (1975), 337-46

Leighly, J. 'Carl Ortwin Sauer, December 24, 1889-July 18, 1975', *Ann. Assoc. Am. Geogr.*, vol 66 (1976), 337-48

Pfeifer, G. 'Carl Ortwin Sauer, 24.12.1889-18.7.1975', *Geogr. Z.*, vol 63 (1975), 161-9

2. PRINCIPAL WORKS OF CARL ORTWIN SAUER

a. Regional Monographs

1916 'Geography of the Upper Illinois Valley and history of development', *Illinois Geol. Surv., Bull.* no 27, 208p.

1920 'The geography of the Ozark Highland of Missouri', *Bull. Geogr. Soc. Chicago*, no 7, 245p.

1927 'Geography of the Pennyroyal', *Kentucky Geol. Surv.* ser 6, vol 25, 249p.

b. Historical Geography

1927 'Lower California Studies. I, Site and Culture at San Fernando de Velicata', *Univ. California Publ. Geogr.*, vol 2, no 9, 271-302, (with Peveril Meigs)

1930 'Pueblo Sites in Southeastern Arizona', *Univ. California Publ. Geogr.*, vol 3, no 7, 415-59, (with Donald Brand)

1931 'Prehistoric settlement of Sonora, with special reference to Cerros de Trincheras', *Univ. California Publ. Geogr.*, vol 5, no 3, 63-148, (with Donald Brand)

1932a 'Aztatlán: prehistoric Mexican Frontier on the Pacific Coast', *Ibero-Americana*, no 1, 92p., (with Donald Brand)

1932b 'The road to Cíbola', *Ibero-Americana*, no 3, 58p.

1934 'The distribution of aboriginal tribes and languages in Northwestern Mexico', *Ibero-Americana*, no 5, 94p.

1935 'Aboriginal population of Northwestern Mexico', *Ibero-Americana*, no 10, 33p.

1937 'The discovery of New Mexico reconsidered', *New Mex. Hist. Rev.*, vol 12, 270-87

1941 'The credibility of the Fray Marcos Account', *New Mex. Hist. Rev.*, vol 16, 233-43

1945 'The relation of man to nature in the Southwest', *Huntington Library Quarterly*, vol 8, 116-26

1948 'Colima of New Spain in the 16th century', *Ibero-Americana*, no 29, 124p.

1960 'Middle America as culture historical location', *Actas del XXXIII Congr. Int. de Americanistas, San José, Costa Rica, 1958*, vol 1, 115-22

1966 *The Early Spanish Main*, Berkeley and Los Angeles, 306p.

1968 *Northern mists*, Berkeley and Los Angeles, 204p.

1971 *Sixteenth century North America: the land and the people as seen by the Europeans*, Berkeley and Los Angeles, 319p.

1977 'Seventeenth century North America', Turtle Island Foundation, Berkeley

c. Cultivation of Plants in Prehistory

1936 'American agricultural origins: a consideration of nature and culture', in *Essays in anthropology Presented to A.L. Kroeber*, Berkeley, 278-97

1947 'Early relations of man to plants', *Geogr. Rev.*, vol 37, 1-25

1950 'Cultivated plants of South and Central America', in *Handbook of South American Indians, Smithson. Inst., Bur. Am. Ethnol., Bull.* 143, vol 6, 319-44

1952 *Agricultural Origins and Dispersals*, Am. Geogr. Soc., New York, 110p.

1959 'Age and area of American cultivated plants', *Actas del XXXIII Congr. Intern. de Americanistas, San José, Costa Rica, 1958*, vol 1, 213-29

d. The Paleogeography of Man

1944 'A geographic sketch of early man in America', *Geogr. Rev.*, vol 34, 529-73

1948 'Environment and culture during the last
      glaciation', *Proc. Am. Philos. Soc.*, vol 92,
      65-77
1957 'The end of the Ice Age and its witnesses',
      Geogr. Rev., vol 47, 29-43
1961 'Sedentary and mobile bent in early man', in
      S.L. Washburn (ed), *Social life of early man*,
      Viking Fund Publ., in *Anthropology*, Chicago,
      no 31, 258-66
1962 'Fire and early man', *Paideuma, Mitt. für
      Kulturkunde*, vol 7, 399-407
1964 'Concerning primeval habitat and habit', in
      Festschr. ad. E. Jensen, München, 513-24
1970 'Plants, animals and man', in R.E. Buchanan,
      E. Jones, and D. McCourt (ed), *Man and his
      habitat*, London, 34-61

e. Geography as an academic discipline
1924 'The survey method in geography and its
      objectives', *Ann. Assoc. Am. Geogr.*, vol 14,
      17-33
1925 'The morphology of landscape', *Univ. California
      Publ. Geogr.*, vol 2, no 2, 19-53
1927 'Recent developments in cultural geography', in
      E.C. Hayes (ed), *Recent developments in the
      social sciences*, New York, 154-212
1941 'Foreword to historical geography', *Ann. Assoc.
      Am. Geogr.*, vol 31, 1-24
1956 'The education of a geographer', *Ann. Assoc. Am.
      Geogr.*, vol 46, 287-99
1966 'On the background of geography in the United
      States', *Heidelberger Geogr. Arb., Festgabe zum
      65. Geburtstag von Gottfried Pfeifer*, vol 15,
      59-71

f. Miscellaneous
1929 'Land Forms in the peninsular range of California
      as developed about Warner's Hot Springs and Mesa
      Grande', *Univ. California Publ. Geogr.*, vol 3,
      no 4, 199-290
1930 'Basin and range forms in the Chiricahua area',
      Univ. California Publ. Geogr., vol 3, no 6,
      339-414
1938 'Theme of plant and animal destruction in
      economic history', *J. Farm Econ.*, vol 20, 765-75
1939 *Man in nature: America before the days of the
      White Man, a first book in geography*, New York,
      273p.
1956 'The agency of man on the earth', in William
      L. Thomas, Jr. (ed), *Man's role in changing the
      face of the earth*, Chicago, 49-69
1963 'Homestead and community on the Middle Border',
      in H.W. Ottoson (ed), *Land use policy in the
      United States*, Lincoln, Nebraska, 65-85
1963 *Land and life: a selection from the writings of
      Carl Ortwin Sauer*, John Leighly (ed), Berkeley
      and Los Angeles, 435p.

*John Leighly is Emeritus Professor of Geography at the
University of California, Berkeley.*

CHRONOLOGICAL TABLE: CARL ORTWIN SAUER

Dates	Life and career	Activities, travel, fieldwork	Publications	Contemporary events and publications
1889	Born, Warrenton, Missouri			
1891				Ratzel: *Anthropogeographie*, vol 2
1899-	In school, Calw (Württemberg)			Schlüter: Bemerkungen zur Siedlungsgeographie (*Geogr. Z.*, 5:65-84)
1903				Department of Geography, University of Chicago, established
1908	B.A., Central Wesleyan College			
1908-1909	Student, Northwestern University			
1909	Student, University of Chicago			Davis: *Geographical Essays*
1911			First work: *Educational opportunities in Chicago*	Semple: *Influences of Geographic Environment*
1913	Married to Lorena Schowengerdt	Map editor, Rand McNally Co., Chicago		
1914-1915	Instructor, State Normal School, Salem, Massachusetts			
1915	Ph.D., University of Chicago; Instructor, University of Michigan			
1916			*Geography of the Upper Illinois Valley*	
1920		Director, University of Michigan Summer Camp, Kentucky	*Geography of the Ozark Highland of Missouri*	
1922	Professor, University of Michigan	Field Director, Michigan Land Economic Survey		Michigan Land Economic Survey established
1923	Professor, University of California			
1925		Began fieldwork in Mexico: almost annual journey to 1945	The morphology of landscape	
1927			*Geography of the Pennyroyal*	

Dates	Life and career	Activities, travel, fieldwork	Publications	Contemporary events and publications
1934		Member, Selection Committee, Guggenheim Memorial Foundation (to 1963); Adviser, Soil Conservation Service	The road to Cíbola	U.S. Soil Conservation Service established
1936			American agricultural origins	
1939			*Man in nature*	
1940	Daly Medal, Am. Geogr. Society	President, Assoc. Am. Geographers		
1941-1942		Travelled in South America: Columbia to Chile	Foreword to historical geography	
1946		In Venezuela		
1949		Mem., Bd. of Visitors, U.S.A.F. Air Univ., travelled in Central and South America		
1951		Bowman Memorial Lecturer, Am. Geogr. Soc.		
1952		In West Indies	*Agricultural origins and dispersals*	
1955	Phil. D. (h.c.), University of Heidelberg	In Europe: Chairman, Symposium on Man's Role in Changing the Face of the Earth; Hon. President, Assoc. Am. Geographers		
1956			The education of a geographer	
1957	Retired; Vega Medal, Svenska Sällsk. f. Antrop. o. Geogr.			
1958	LL.D., University of Syracuse, New York			
1959	Humboldt Medal, Gesellsch. f. Erdk. zu Berlin	Travelled in Europe		
1960	LL.D., University of California			
1965	LL.D., University of Glasgow	Visit to Europe		
1966			*The early Spanish Main*	
1968			*Northern mists*	

Dates	Life and career	Activities, travel, fieldwork	Publications	Contemporary events and publications
1971			Sixteenth century North America	
1975	Victoria Medal, R. Geogr. Soc., London; died at Berkeley, California			

Mary Somerville
1780-1872

MARGUERITA OUGHTON

By permission of The Scottish National Portrait Gallery

Mary Somerville was a woman of brilliant intellect whose claim to influence in the development of geographical thought derives from her scientific and geographical writing.   She lived in an era when women were neither expected nor encouraged to participate in the worlds of science or of learning.

1. *EDUCATION, LIFE AND WORK*

Born Mary Fairfax, at Jedburgh, on 26 December 1780, she came from well-connected families.   Her father was a distinguished English naval officer, Vice-Admiral Sir William Fairfax (1739-1813), who was knighted after the Battle of Camperdown (1797). From her mother, a daughter of Samuel Charters, Solicitor of Customs for Scotland, she got her Scottish background and upbringing.

Her formal education was minimal, and it was through her own persistence and in spite of family discouragement that she developed her flair for mathematics and pursued the interests stimulated by her innate curiosity, her powers of observation and her intellect.   From her home at Burntisland, Fife, in early widowhood after a brief first marriage, she embarked in 1807 on serious self-education in mathematics and earned respect in the brilliant intellectual and scientific circles of that time in Edinburgh.

A second marriage, in 1812, to a first cousin, Dr William Somerville, led to residence in London after 1816 and to close involvement in the capital's scientific, literary and social circles.   William Somerville was elected to the Royal Society in 1817.

He enthusiastically encouraged his wife's writing in astronomy and mathematics.   The Somervilles associated with the leading scientific and philosophical thinkers not only of Britain but also of Europe.

In 1838 they and their two surviving daughters left England to settle eventually in Italy, because of a change both in their private fortune and in Dr Somerville's health.   Mary Somerville continued her assiduous reading, writing and correspondence with leading scientists and explorers until her death in Naples on 29 November 1872.

Mary Somerville seems to have had a personality which, while she asserted her inborn faculties and established herself as a scientific authority, enabled her at the same time to adapt her life without bitterness to the restraints and frustrations inevitable for a learned woman in Victorian times.   As well as her gifts for organising scientific information and for lucid writing, she was endowed with modesty, feminine charm and the ability to achieve cultural and domestic accomplishments.

2. *SCIENTIFIC IDEAS AND GEOGRAPHICAL THOUGHT*

Mary Somerville's contribution to nineteenth-century geography was essentially through two of her books: *On the connexion of the physical sciences* and *Physical geography*.   In both of these she proved herself to be a supreme expositor, drawing together the writing and thinking of a wide field of scientific writers and thinkers, including geologists and explorers.   From her reading and her correspondence with a large number of authorities, she assimilated a vast range of material, in the editing of which she established the relationships between new discoveries in the fields of science and geographical exploration. By constant revision, she kept her texts up to date with current literature.

Physical sciences has its origin in the 'Preliminary dissertation' with which she prefaced an earlier book, *The mechanism of the heavens* (1831) which was a translation-cum-exposition of La Place's *La mécanisme céleste* (1799-1825). La Place was reported to have said that Mary Somerville was the only woman who understood his works. First published in 1834, *Physical sciences* touched on the planetary system, earth history, tides, properties of the atmosphere, the distribution of temperature and of plants and animals, ocean currents and climate. It ran through ten editions up to 1877, the first five being published by 1840. All but the last were revised by the author.

Study of her works suggests that Mary Somerville started collecting material for *Physical geography* during the 1830s, in the course of revising *Physical sciences*. She doubtless drew on the sources offered by the new journal of the Royal Geographical Society (established in 1830). This period saw the emergence of 'physical geography' to which the seventh edition of the *Encyclopaedia Britannica* (1834) devoted a separate article. Moreover, Mary Somerville had great admiration for and used the work of geologists like Sir James Hall, Sir Roderick Murchison and Sir Charles Lyell, the last of whom she had known since their youth. She was herself profoundly interested in geographical exploration and description.

Although the first volume of *Physical geography* was ready by 1842, the whole book was not published until 1848. Could it have been brought out earlier, it would probably have achieved longer lasting fame, foreshadowing Humboldt's great *Kosmos* (of which the first volumes were published in 1845) and long preceding other physical geography texts which appeared soon after 1848. It ran through seven editions, of which the two last were revised by H.W. Bates.

Mary Somerville differs from the authors of previous English geographical texts by being concerned not merely with the distribution of phenomena, country by country, but also with the causes underlying those distributions and with relationships between various physical phenomena. She also adopted a systematic approach on continental or zonal bases, in contrast to the political approach, constrained by national boundaries, which was commonly used in geographical textbooks. In this sense her writing foreshadowed the regional approach in textbooks.

Physical geography carried neither maps nor illustrations but in the second edition (1849) she referred readers to A.K. Johnston's *Physical atlas*. This was a specially prepared small edition of his *Physical atlas of natural phenomena* which had appeared in 1848. A.H. Petermann had been employed in the preparation of this successful and widely acclaimed work, which derived much from Berghaus's famous atlas. Mrs Somerville was both very much up to date in her sources and references and generous in her acknowledgement of them.

Although her text bears witness to her devout religious beliefs, her book appeared during the years leading up to Darwin's publication of *The origin of species* (1859) and during a period of intense hostility towards evolutionary ideas. While she concealed none of the implications of scientific discovery for evolutionary theories, she avoided explicit reference to them and, according to a letter from her publisher, seems to have been anxious, probably through the influence of the Duke of Argyll and John Herschel, that 'no Darwinism' would appear in the sixth edition of *Physical geography* (1870) which was to be revised for her by H.W. Bates, assistant secretary at the Royal Geographical Society and an adherent of the Darwinian hypothesis. An anomalous sociological and philosophical review of current affairs was appended to the final chapter ('The distribution, condition and future prospects of the Human Race') from the first to the fifth editions, curiously revealing Mrs Somerville's attitudes to contemporary social and national conditions in Britain and Europe. This section was eliminated by Bates in his revision of the sixth edition.

The revisions of *Physical geography* over a period of thirty years reflect the development of geographical knowledge and discoveries during the middle half of the nineteenth century, within a compass and a style of writing that was convenient for general reading but which required an educated approach. In this sense it differed from contemporary geographical textbooks, such as those of William Hughes, which were manuals of location and description, intended mainly for memorizing.

3. INFLUENCE AND SPREAD OF IDEAS

Acknowledged as an authoritative scientific writer from the 1820s onwards, and with the support of her publisher, John Murray -- a reputable publisher with close connections with the literary and scientific world -- Mary Somerville was assured of acceptance in learned circles. Commenting on her manuscript of *The mechanism of the heavens*, Sir John Herschel wrote 'Go on thus, and you will leave a memorial of no common kind to posterity'. Soon after the publication of this work, Dr W. Whewell and Professor George Peacock introduced it into their courses at Cambridge; it was used as a textbook for almost a century. Of *Physical sciences* James Clark Maxwell wrote that it was one of those 'suggestive books which put into definite intelligible and communicable form the guiding ideas that are already working in the minds of men of science, so as to lead them to discoveries but which they cannot yet shape into a definite statement'. It was translated into German and Italian and was pirated in American editions.

These two books brought Mary Somerville much honour. In 1833 the Royal Astronomical Society named her and the elderly Caroline Herschel as its first female honorary members. The Royal Society commissioned from the sculptor Chantrey a bust of Mrs Somerville, to be placed in the Society's Great Hall (where it still stands). Her books earned her a Civil Pension of £200 annually (raised to £300 in 1837).

Physical geography won instant approbation -- from Humboldt and Herschel among others -- and was clearly accepted as an authoritative work.   It appeared at a significant time for physical geography in the universities of Oxford and Cambridge. In 1848 the Natural Sciences Tripos was established at Cambridge which included geology, of which it may be assumed that physical geography was part.   At Oxford physical geography was recognized in 1850 as a separate optional subject in the new School of Natural Science, along with geology, botany, zoology and mineralogy.

Mary Somerville's book appeared on the book lists at Oxford, and also at Manchester from the 1870s to the 1890s; and in the list of reference works compiled from the Royal Geographical Society by J. Scott Keltie in 1886.   It was cited alongside Humboldt's *Kosmos* in the *Edinburgh Review* notice of Keith Johnston's *Physical atlas* in 1849 and provided a source reference for several authors of manuals, sometimes acknowledged, often without any acknowledgement.   Professor D.T. Ansted, of King's College, London, acknowledged his use of it in the preparation of Part 3, 'Physical geography', in C.G. Nicolay's *A manual of geographical science* (1852).   Sir John Herschel who wrote the article 'Physical geography' for the *Encyclopaedia Britannica* (8th edition, 1853-60), singled it out, alongside Johnston's *Physical atlas* and Ansted's work, as special authorities he had consulted.   Mary Somerville's book, from the first edition, had been dedicated to Herschel, as a mark of his long friendship and encouragement of her work.

All was not praise, however.   Mary Somerville was publicly denounced in the House of Commons after *The mechanism of the heavens* appeared; and after the publication of *Physical sciences*, she was mentioned by Dean Cockburn from York Minster in his denunciation of the new geology.   The eighteenth-century struggle between science and religion was still far from resolved.   Professor Elizabeth Patterson suggests that these events may in part have lain behind her evasion of explicit reference to Darwinian hypotheses in her writings.

Nevertheless numerous honours, in many countries, were bestowed on Mary Somerville.   Three of these, in 1869, towards the end of her long life, were the Victoria Gold (Patron's) Medal of the Royal Geographical Society, London (which did not admit women as members until 1913), the Victor Emmanuel Gold Medal (the first to be awarded by the Geographical Society of Florence) and election to the American Philosophical Society.

The best insight into Mary Somerville's life and work so far published is in articles published by Professor Elizabeth Patterson (1969 and 1974); J.N.L. Baker (1948) examined her place in the development of geography in England.   Her own *Personal recollections from early life to old age* (1873) was edited by her daughter and published posthumously by John Murray.   An unedited draft, part of the Somerville papers in the Bodleian Library, Oxford, awaits research and editing.   Her obituary notice in the London *Morning Post* recognized her as the

'Queen of Nineteenth-Century Science'.   Her work is now little remembered but her name is commemorated in that of Somerville College, Oxford, called after her posthumously on its establishment in 1879.

Bibliography and Sources

1. REFERENCES ON MARY SOMERVILLE

Somerville, Martha (ed) *Personal recollections: from early life to old age of Mary Somerville, with selections from her correspondence*, London, 1873, 377p.

Baker, J.N.L. 'Mary Somerville and geography in England', *Geogr. J.*, vol 111 (1948), 207-22

Patterson, Elizabeth C. 'Mary Somerville', *Br. J. for the Hist. of Sci.*, vol 4 (1968-9), 311-39

Patterson, Elizabeth C. 'The case of Mary Somerville: an aspect of nineteenth-century science', *Proc. Am. Philos. Soc.*, vol 118 (1974), 269-75

Sanderson, Marie 'Mary Somerville. Her work in physical geography', *Geogr. Rev.*, vol 64 (1974), 410-20

Dictionary of Scientific Biography, vol II (1975), 521-5

2. WORKS BY MARY SOMERVILLE

1826 'On the magnetizing power of the more refrangible solar rays', *Philos. Trans. (R. Soc.)*, vol 96, 132

1831 *The mechanism of the heavens*, London, 621p.

1832 *A preliminary dissertation on the mechanism of the heavens* (Preface to the previous reference, reprinted and sold as a separate volume), London and Philadelphia

1834 *On the connexion of the physical sciences*, London, 356p. 10 editions up to 1877

1835 'Astronomy -- the Comet', *Q. Rev.*, vol 105, 195-233

1836 'Experiments on the transmission of chemical rays of the solar spectrum across different media', *C.R. Acad. des Sci.*, vol 3, 473-6

1843-50 'On the action of the rays of the spectrum on vegetable juices', extract from letter to Sir John Herschel, Rome 20 Sept. 1845, *Abstr. Philos. Trans. (R. Soc.)*, vol 5, 569

1848 *Physical geography*, London, 381p. Six further editions with revisions to 1877

1869 *On molecular and microscopic science*, London, 1869, 2 vols

3. UNPUBLISHED SOURCES

The Somerville Collection of MSS, papers, letters, documents, diplomas and memorabilia is deposited in the Bodleian Library, Oxford.

Marguerita Oughton was a Research Associate in the University of Manchester from 1975-1976 and, with T.W. Freeman and Philippe Pinchemel, edited the first volume in this series.   She is now Librarian and Map Curator in the School of Geography, Manchester University.

Paweł Edmund Strzelecki
1797–1873

JÓZEF BABICZ
WACŁAW SLABCZYŃSKI AND
THOMAS G. VALLANCE

Strzelecki was a traveller and explorer of marked individuality and immense enthusiasm, whose work -- especially in Australia -- was of lasting significance.   Unlike many explorers he was a geographer of sufficient merit to become a figure of note in the history of the subject.

1. LIFE, EDUCATION AND WORK

The son of Franciszek and Anna (née Raczyńska), Strzelecki came from a once famous landed family in central Poland and both his parents were of noble birth.   His mother died in 1810 when he was thirteen years old, and he was taken to Warsaw by relatives, where he went to school from 1810-14.   He lived with a family favouring French revolutionary ideas.   In the period of the Napoleonic Wars the creation of the Duchy of Warsaw from 1807-15 was in effect an expression of a relatively free Poland, not however to be realized for more than a century.   Though the record of Strzelecki's life as a young man is obscure, he lived for some time in Cracow and travelled round the country.   He seems also to have spent some time abroad, visiting Italy and Switzerland.   From 1825-9, he administered the vast estates of Prince F.K. Sapieha at Dereczyn in Belorussia.   When his employer died he received a large sum, 12,000 ducats, as compensation for the loss of his position.   This enabled him to pursue his scientific interests.   Although of gentle birth, he was opposed to the parochial and conformist outlook of the gentry in the area of Poland annexed by Prussia and remained faithful to the romantic and patriotic ideas of his youth.   Finally, he took part in the national uprising against Russia in 1830 and with its failure left Poland with the numerous emigrés and settled in England.

During his first travels in England, Wales and Scotland, he was interested in agriculture and animal husbandry, and he steadily developed a liking for the natural sciences, notably geology.   It was at this time that he acquired the considerable knowledge of geology and natural sciences demonstrated in his later travels.   He appears to have gained this knowledge by self-education and personal contacts with scientists and scientific institutions in England.

On 8 June 1834, he set out from Liverpool for New York, on the first stage of his projected journey round the world.   He went to Niagara Falls, then to Canada, and became interested in geological prospecting and in the life of the Huron Indians.   In the Great Lakes area he found copper deposits and reported on them to the British government.   He also made a journey from New York to the West Indies and Mexico and returned to the United States by way of New Orleans, whence he travelled up the Mississippi river. While in the United States he was much concerned with the impoverishment of the soil and studied the amount of gluten in wheat grown in Maryland and Virginia. He also investigated the ethnography of the Seminoles. At various times during his travels he came across Poles who had fled to America after the 1830 rising. Many of them were in poor circumstances and he helped them with money and also intervened on their behalf with influential Americans likely to be able to assist them.

In the years 1836-8 he travelled through South America, mainly in Brazil, Argentina, Paraguay and along the west coast, particularly in Chile, and then returned to Mexico and California. His varied studies included meteorological investigations of sub-tropical areas, volcanoes in Mexico, the extraction of gold and silver, the analysis of soils and their crops, and the state of rural economy. And to all this he added ethnographical, historical and, occasionally, archival investigations.

In the middle of 1838 he travelled westwards across the Pacific on the British ship 'Fly' to the Marquesas islands and next to the Hawaiian archipelago, where his vulcanological investigations were published in the *Hawaiian Spectator* of 1838 and reprinted in other journals. Travelling further to other island groups he was a careful observer of the indigenous people and is reputed to have given helpful advice on an appropriate judicial system for Tahiti. In 1839 he reached New Zealand where he studied the life of the Maoris and conferred with their chieftains. He also investigated the nature and geology of the land he intended to describe at some later time.

Finally, on 25 April 1839, Strzelecki arrived in Australia with the avowed aim of investigating the natural and mineral resources of the continent. His enterprise was welcomed by the Governor of New South Wales, Sir George Gipps, and his purpose was well described in his report sent by the Governor in 1840 to the British parliament:

> I was particularly anxious to ascertain the course of the dividing range, its configuration, elevation, ramifications, connexion with the subordinate protuberances and the nature of the simple or compound minerals removed by its rise from the entrails of the earth to the present surface.

During his explorations he climbed the highest peaks, determined their altitudes, gave them names and followed the course of mountain streams. He also carried out investigations on the climate, earth magnetism, flora and fauna (present and fossil), and on the possibilities of growing crops, raising animals and finally on the life of the Aborigines.

All this was to be inserted in a monograph on this part of the world, for which he made copious notes in his diary. He also collected samples of rock and made chemical analyses of soils. His geological and mineralogical prospecting yielded scientifically interesting results. For instance he found occurrences of iron oxides, of iron pyrites and of kaolin and made useful observations on coal. He found many other minerals including gold between Wellington and Bathurst on the western side of the Blue Mountains. Governor Gipps, however, kept this last find secret; exploitation began much later, in 1851.

Strzelecki's second expedition began on 21 December 1839. With seven men and six pack horses he went to the Great Dividing Range, explored several valleys and basins and crossed to the western side, where he was joined at Ellerslie pastoral station by his friend, J. Macarthur, a nineteen-year-old English boy J. Riley, a native guide and a servant. Having been supplied with notes and other materials by P. King, proprietor of the grazing property of Dutzon, Strzelecki explored and mapped the country between the rivers Murrumbidgee and Murray, following several streams to their sources. Finally he travelled up the Murray to the highest mountain in Australia, which he reached on 12 February 1840 and named Mount Kosciusko, as a tribute to his countryman who had fought for the independence of America and Poland.

a. The discovery of Gippsland

Proceeding to the south over various valleys and basins, Strzelecki, in the vicinity of Lake Omeo and Mount Tambo, crossed the Great Dividing Range, moving to the south-east. Turning next to the south-west he entered an unknown land which, because of its rich pastures and resources, was to become an area important to the Australian economy. Strzelecki named it in honour of the Governor Gipps. On the western part of this newly discovered country however the party encountered an almost impenetrable scrub. After many tribulations, Strzelecki and his party, utterly exhausted, arrived at Melbourne in May 1840. Here Strzelecki suggested the immediate exploitation of Gippsland and won the support of the Commissioner of Crown Lands at Port Philip, H.F. Gisborne.

b. Tasmanian exploration

Tasmanian exploration followed, and in July 1840 Strzelecki was warmly welcomed at Launceston by the island's Lieutenant-Governor, Sir John Franklin, the Polar explorer. There were three major explorations in Tasmania. Finally some months were spent in collecting and arranging materials before Strzelecki left Launceston for Sydney on 29 September 1842. In the next few months he visited the Hunter Valley and adjacent parts of New South Wales.

c. Return to England

On 22 May 1843, after almost four years there, Strzelecki left Australia. On the way to England he spent four months in the Far East, visiting the Malayan Archipelago, Malaysia, China and India. On 24 October he reached London.

Nine years spent in travels and explorations left Strzelecki impoverished and in poor health. He had received much kindness from Franklin and from officers of the British Navy. He had also accumulated a mass of material for future work in his notes and boxes of specimens. In 1845 *Physical description of New South Wales and Van Diemen's Land* was published with the help of the people of Tasmania who, with a thankful address for his work in Tasmania, forwarded him a subscription of £400 for this purpose. In the following year he was given the Founder's Medal of the Royal Geographical Society. In 1845 he became a naturalized British citizen, in 1853 a fellow of the Royal Society and a member of the Council of the Royal Geographical Society.

In the meantime, the Irish Famine had shocked an England little aware of such disasters and with no experience of relief work. The British Association for the Relief of the Extreme Distress in Ireland and Scotland engaged Strzelecki's honorary services, first as a field officer in the west of Ireland and then as its main representative in the Dublin headquarters. In his reports the all-too-familiar story of the Irish national tragedy was told. On 21 November 1848 he was appointed by Queen Victoria to the Order of the Bath in recognition of his services. He maintained his links with the Royal Geographical Society and helped to raise funds for the expedition led in the years 1857-9 by Captain L.F. McClintock in search of John Franklin, lost in the Arctic.

Australian interests were also maintained. For instance he outlined a scheme to irrigate the southern portion of New South Wales that was, in the opinion of Professor S.H. Roberts of Sydney University, 'altogether the largest and most practical of the many colonisation ventures then contemplated in Australia'. In 1853-7 Strzelecki acted as Chairman and Director for the Peel River Land and Mineral Company. He was always a valued government adviser on Australian affairs but never received the costs of his Australian expeditions despite the promises of Governor Gipps that they would be forthcoming. This hope appears to have perished with the Governor's death. Nevertheless, the years of his unselfish scientific and social works, of exertion and perilous explorations had been rewarding. During his London period, covering thirty years from 1843, he maintained friendly relations with many of the most eminent scientific men of the day, including Charles Darwin, Roderick Murchison, Charles Lyell, Sir John Herschel. Many of them praised his work and regarded him with esteem and even admiration. He was in close contact with the bereaved Lady Franklin, the renowned philanthropist Florence Nightingale and he also maintained friendly relations with W.E. Gladstone, several times Prime Minister, who in 1869 secured Strzelecki's advance to the honour of Knighthood (K.C.M.G.). Strzelecki died in London on 7 October 1873.

2. SCIENTIFIC IDEAS AND GEOGRAPHICAL THOUGHT

Although Strzelecki never published the results of his American studies, the fragmentary records show that he was interested in the climate, geology, soils, rural economy and population of this part of the world too. On the whole his explorations here and elsewhere were the background for a traditional regional geography, based on the collection of facts, their interpretation, correlation and synthesis. He also viewed phenomena comparatively, noting similarities between these seen in various parts of the world and, as he himself said, capturing any identity or analogy between them.

His main work, the *Physical description of New South Wales and Van Diemen's Land*, was based on his own observations and on wide reading. Its lasting importance calls for some analysis. Of the eight sections, six deal with the natural environment, one with the aboriginal inhabitants and the last with agriculture. Section 1 -- Marine and Land Surveys (pp.1-46) summarizes previous work and introduces Strzelecki's own observations on topographical features, their distribution, altitudes, and so on. The second section -- Terrestrial Magnetism (pp.47-50) records magnetic variation determined at forty-six stations. Section 3 -- Geology and Mineralogy (pp.51-158) is in itself a landmark in Australian science. It is based largely on the explorer's own surveys and collections. The treatment, in terms of four epochs of formation, offered a novel time-classification for Australian geology. Its basis may have been the old-fashioned one of

lithological character but Strzelecki was clearly aware of recent developments in European thought and was able to apply this sophistication. Accompanying the book are reduced versions of the geological sections and map prepared by Strzelecki. The original map, on a scale of about ¼ inch to 1 mile, is preserved in the collection of the Institute of Geological Sciences, London. The fourth section (pp.159-240) deals with what Strzelecki termed Climatology. In his view meteorology ranked next in importance to geology among the branches of physical geography and the section is a dense record of observations related to the atmosphere. With his friend Captain (later Rear-Admiral) P.P. King, F.R.S., Strzelecki must be accounted a pioneer of Australian meteorology. Sections 5 -- Botany (pp. 241-58), and 6 -- Zoology (pp.259-332) introduce a discussion of both fossil and recent flora and fauna respectively. Here Strzelecki was supported by specialists in London to whom he submitted appropriate parts of his collections. The fossil plants and animals, some elegantly illustrated in the book, are roughly linked to the geological section though it is evident Strzelecki had not fully conformed to the advice of the palaeontologists, J. Morris and W. Lonsdale. In the seventh section (pp.333-56), the Aborigines are discussed in terms of their origins, their physical characters, language, beliefs and customs. The influence of European settlement on native Australian communities interested Strzelecki who was able to draw on experience of analogous cases in other countries. He concluded that the fertility of the Aborigines was then diminishing and adduced various physiological arguments to support what became known as the 'Strzelecki law'. Charles Darwin mentioned it, but without comment, in his work. The final Section 8, 'Agriculture' (pp.357-462) deals with the current state of Australian agriculture and its future prospects. Again the treatment is systematic -- nature of soils, utilization of land, types of farming and grazing activities and similar topics. Strzelecki examines likely ways of improving crops such as irrigation and soil fertilization. The author was clearly possessed of a close familiarity with agricultural practice in Europe and with agricultural economics.

Finally, in *Gold and Silver* (London, 1856), a supplement to his *Physical description...*, Strzelecki gives his story of the discovery of gold in Australia. Apart from a chronology of events, in itself a useful study, he courteously disputes the principle of priority in any individual discovery. He mentions that Sir Roderick Murchison, when he saw his geological sections and samples and compared them with similar finds in the Urals, came to the conclusion that Australia was an auriferous country long before the so called 'discovery' of E.H. Hargraves.

3. INFLUENCE AND ACHIEVEMENT

Strzelecki made notable contributions to the exploration of Australia, not merely as a discoverer but as one of the first to examine particular aspects of the country's physical geography. For many years his *Physical description* was the most complete account available of any substantial part of Australia and, unlike earlier geographies, the book was based on the author's own travels. From the economic point of view, there is little doubt that Strzelecki's discoveries in Gippsland were of great value to Australia. Long after he left Australia, Strzelecki continued to be a respected source of advice to colonial authorities, both government and mercantile.

Strzelecki was essentially a pragmatic field scientist who, like many of his contemporaries, saw scientific explorations as a source of benefit to mankind. That, to him, was an important aspect of practical philanthropy. Certainly the economic development of Australia in the second half of the nineteenth century lay on foundations outlined in his book of 1845. It is not surprising that his name appears attached to many places on the map of Australia and is commemorated in the names of certain elements of the Australian flora and fauna, both fossil and recent.

Bibliography and Sources

The first scientific estimations of Strzelecki's activities are to be found in the reviews of his books:

Brewster, D. 'Physical description of New South Wales', *North Br. Rev.*, vol 4 (1846), 281-312

Buckland, W.L. 'Physical description of New South Wales', *Q. Rev.*, vol 76 (1846), 488-521

Perkins, J.H. 'A glimpse of Australia', *North Am. Rev.*, vol 25 (1850), 196-7

Works on Strzelecki include:

'Sir Paul Edmund de Strzelecki', *Proc. R. Geogr. Soc.*, vol 18 (1873-4), 518-23 (in annual address by Sir H. Bartle Frere)

Zmichowska N. 'O Pawle Edmundzie Strzeleckim (P.E. Strzelecki in the memories of his family and friends)', *Ateneum*, vol 1 (1876), 387-424 and 538-96

Havard, W.C. 'Sir Paul Edmund Strzelecki', *J. Proc. R. Aust. Hist. Soc.*, vol 26 (1940), 20-47

Dowd, B.T. 'The Cartography of Mount Kościuszko', *op.cit.*, 97-107

The *Life of Sir Paul Edmund Strzelecki, 1796-1873*, was published by the Strzelecki Committee to commemorate the seventieth anniversay of Strzelecki's death in London (1943), 32p. and illustrations

Rawson, G. *The Count, a life of Sir Paul Edmund Strzelecki KCMG, explorer and scientist*, London (1954), 214p. with illustrations and maps.

Wacław Słabczyński *Paweł Edmund Strzelecki, Podróże -- odkrycia -- prace (P.E. Strzelecki, journeys -- discoveries -- works)*, Warsaw (1957), 327p. with maps, illustrations and index. This book, based thoroughly on sources, comprises also a bibliography of *c.* 100 minor works and statements on Strzelecki. It opens a three-volume series on Strzelecki, of which vol 2 is a translation into Polish of the work *Physical description of New South Wales and van Diemen's Land* (London 1845, 462p.); vol 3 bears the title: *Paweł Edmund Strzelecki, selected works*, collected and edited by Wacław Słabczyński, Warsaw (1960), 270p. The selection comprises nearly all known Strzelecki's original works and also some of his correspondence with J. Franklin and C. Darwin.

Another work is:

Clews, Hugh Powell G. *Strzelecki's ascent of Mount Kościuszko 1840*, Melbourne (1973).

Manuscripts not yet used include a geological map in the Geological Museum, London, and some correspondence in the Royal Geographical Society.

Professor Józef Babicz is head of the department of the History of Natural Sciences in the Institute of the History of Science and Technology of the Polish Academy of Sciences: Dr. Wacław Słabczyński is Assistant Librarian (Emeritus) of the National Library, Warsaw. Thomas G. Vallance is Professor of Petrology in the University of Sydney.

PLACE NAMES

Eight places have been named after Strzelecki, all except one of them in Australia.

The town of *Strzelecki*, 30° 20' S, 145° 54' E, is in Gippsland, 80 km southeast of Melbourne at the foot of Mount Trafalgar, *Victorian Municipal Directory* 1911, 563

Strzelecki Creek (29° 15' S, 139° 59' E), is an intermittent river in Southern Australia to the east of lake Eyre. From lake Blanche it descends to join the Cooper river near Innamincka. This feature was named by the explorer Charles Sturt: see his *Narrative of the expedition into central Australia during the years 1844, 1845, 1846*

Strzelecki Pass, 36° 23' S, 148° 18' E, 1980 m high, is in the Great Dividing Range, here known as the Australian Alps. It is 8 km north of Mount Kosciusko, in the vicinity of Mount Townsend, west of Blue Lake and northwest of lake Albina, Mitchell, E., *Australia's alps*, Sydney and London, 1946

Strzelecki Peaks, 40° 12' S, 148° 05' E, is a mountain range in the southwest part of Flinders island, Bass strait. It has the highest point of the island, at 777 m and was named by Captain J.L. Stokes of the *Beagle* in 1842. Stokes, J.L., *Discoveries in Australia* II, 1846, 419

Strzelecki's Ranges (or Hills), 38° 30' S, 146° 30' E, covers an isolated mountain range in the district of Western Port, Western Gippsland, Victoria, with a maximum altitude of 643 m. *Aust. Geogr.*, vol 8/2 (1961), 56

Mount Strzelecki, 21° 09' 04" S, 133° 52' 40" E, is a mountain in the Northern Territory, at the northwestern extremity of Crawford range, *c.* 25 km west of Stuart Highway and 300 km north of Alice Springs. It was named on 23 May 1860 by John McDouall Stuart during his fourth expedition

South Strzelecki, 38° 21' S, 145° 53' E, is a locality in Gippsland, Victoria

The Canadian example is:

Strzelecki Harbour, 75° 15' N, 96° 29' W, in the southeast of Prince of Wales island, Franklin district. It was named by F.L. McClintock who was in charge of the Franklin rescue expedition.

CHRONOLOGICAL TABLE: PAWEL EDMUND STRZELECKI

Dates	Life and career	Activities, travel, fieldwork	Publications	Contemporary events and publications
1797	Born at Gluszyna, Poland			Duchy of Warsaw, 1807-15
1810-15	Secondary education in Warsaw			
1815-25	Lived in Cracow and in country houses of relatives	Travel in Poland and abroad (Switzerland and Italy)		
1825-9		In charge of F.K. Sapieha's estates, Dereczyn, Belorussia		
1830	Involved in Polish national rising and migrated to England	Travelled extensively in England, Wales and Scotland		National Rising in Poland
1834-6	World travel began	Travelled in U.S.A., Canada, Mexico, Cuba		
1836-9		In South America and Oceania (Marquesas, Hawaii, Tahiti)	Vulcanological investigation in *Hawaiian Spectator*	
1839-43	Began active exploration in Australia, including Tasmania	Arrived in New Zealand but stayed only for a few weeks, reaching Australia on April 25		Strzelecki's explorations reported in *House of Commons Papers*, vol 17, 11-47 and in Australian publications
1843	Left Australia for London	Travelled in Asia and Africa, May to September but settled permanently in London	*Geological map of New South Wales and Van Diemen's Land* (manuscript in Geological Museum, London)	
1845	Became a naturalized British citizen		*Physical description of New South Wales and Van Diemen's Land*, London	Famine in Ireland
1846	Founder's Medal Royal Geographical Society		'The volcano of Kirauea, Sandwich islands', *Tasmanian J. Nat. Sci.*, 2, 12-41	
1847-9		Worked on famine relief with the British Association for the Relief of the Extreme Distress in Ireland and Scotland		
1848	Order of the Bath			
1849			*Report on the condition of Ireland*	
1853	Became member of Council of Royal Geographical Society. Elected member of the Royal Society	Chairman and Director, Peel River Land and Mineral Company (1853-7)		

Dates	Life and career	Activities, travel, fieldwork	Publications	Contemporary events and publications
1856			*Gold and silver*, supplement to *Physical description of New South Wales and Van Diemen's Land* (1845)	
1860	Honorary degree of D.C.L., Oxford University			
1869	Knighthood conferred			
1873	Died in London, 6 December			

Camille Vallaux
1870-1945

F. CARRÉ

A French geographer who wrote voluminously, Camille Vallaux is now little known. His writings were varied, and deal largely with the oceanic regions of northwest Europe, oceanography and marine hydrology. His interests -- and some of his works -- also included political geography and the theory of geography. But despite his devoted work he never acquired the renown of his contemporaries such as Brunhes, de Martonne or Demangeon. This could be variously explained as due to a dull personality, to the lack of significance in his work, or merely to the relative obscurity of his professional life.

1. EDUCATION, LIFE AND WORK

a. The geographical background: Brittany and coasts
Born in humble circumstances in 'a small region of old France, the Vendômois', as he himself said, he was recognized at his school as a boy of unusual intelligence. All through life he was grateful for the work of his school teachers, and especially for his admission to the famous Ecole Normale Supérieure in 1891, along with the future statesman Edouard Herriot and also with Maurice Zimmerman, who became a notable geographer. At the Ecole he studied under Vidal de la Blache, then in his late forties, and became acquainted with Jean Brunhes, older than himself and also with Albert Demangeon and Emmanuel de Martonne, both of whom entered in 1892. Having passed the competitive *agrégation* examination in 1895 he became a teacher at the lycée in Pontivy (Morbihan), from which he moved to Brest in 1896. His concern with

marine geography was stimulated by his years in Brittany but his initial research was historical and his first book, of 1899, was *Les campagnes des armées françaises (1792-1815)*.

In 1901 he left the Lycée at Brest for a post in geography at the Naval College, then housed in a training ship in the roadstead at Brest. Having now turned to geography he asked Vidal de la Blache to be his supervisor for a thesis on Lower Brittany. Not surprisingly his teaching work for naval cadets gave him an increasing interest in the study of shores, especially those of the Atlantic coasts. Summer courses organized for his pupils, on several of which he was present, included trips to the coasts of Scotland, Norway and the Baltic. He also studied the coasts of the Channel Islands and Cornwall. He read the literature in English and German on both areas, where he spent a considerable time ashore. He was also well acquainted with Norway and lectured on all these areas. In his reading he became an admirer of the work of F. Ratzel and his disciples, though he did not accept their views.

Having become a Docteur ès Lettres in 1907, he remained at Brest and acquired a property, Ty Dréo, at Le Relecq Kerhuon, near the Brest roadstead by then, to him, a much-loved landscape. There, in sight of the Elorn, he enjoyed writing in his library surrounded by his family. Most of his writing was done at Kerhuon and after his removal to Paris in 1913 he returned to his Brittany home for long visits every year. On his retirement in 1932 he made it his permanent home, and there he wrote papers until his last years.

b. *New researches and orientations*

Before the First World War he developed a strong interest in political and social geography, partly through the work of F. Ratzel and partly through his friendship with Jean Brunhes, who spent several holidays in Brittany.   This led to the publication of two volumes of the *Encyclopédie scientifique*, in 1908 and 1911.   A humanist by deep conviction as well as by his education, he was saddened by the War and by the frontier problems of the time.   His concern was seen in the articles he wrote on Germany, on devastated cities and on the League of Nations.   After the war, from 1919 to 1932, he taught history at the Janson de Sailly lycée and geography at the Ecole des Hautes Etudes Commerciales.   In this period of work in Paris his long reflection on the theory and methodology of geography bore fruit in his major work, published in 1925, *Les sciences géographiques*.

   He retained his link with the Naval college, for which he acted as an external examiner. There too his earlier works on maritime geography were still respected and he became the chairman of the Ligue Maritime et Coloniale and the compiler of the oceanographical section -- first of *La géographie*, the review of the Société de géographie, and later of the *Annales de géographie*. He was turning increasingly to oceanography, and mainly to marine hydrology on which he wrote in specialist journals: the *Revue hydrographique*, the *Bulletin de l'Institut Océanographique*, and in his large book *Géographie générale des mers*, which took eight years to produce.   While in Paris he made many contributions to the various societies of which he was a member, and especially to meetings at the Institut Océanographique.

c. *Personality and ideas*

Vallaux was a dedicated scholar and teacher like many academic figures of his time.   He was a man of austere and neat appearance, of firm principles, apt to be uncompromising and even irascible when it seemed that scientific truth was in danger. Immensely loyal to his university and his teachers, he impressed his pupils by his lively and enthusiastic teaching, marked by authority and a broad cultural understanding.   His belief in science was, like that of many teachers during the period of the Third Republic, based on atheistic rationalism.   His opinions were expressed vigorously and clearly, at times with more zeal than discretion for there was an element of rashness in his temperament.   If he thought a scientific position to be wrong he fought it, as his controversy with the oceanographer Le Danois shows: he was equally rigid in his opposition to projects such as the Trans-Sahara railway, which he called 'an unproductive extravagance' (*Mercure de France*, 1 March 1924).   His strongly-held views on science made him a valued contributor to various journals rarely penetrated by geographers, such as the *Revue générale des sciences* and the *Revue scientifique*. Despite the literary character of his education, he wished to see geography regarded as essentially a science.

2. *SCIENTIFIC IDEAS AND GEOGRAPHICAL THOUGHT*

A. *Contributions to Science*

His contribution to science lay in three main areas: the maritime countries of northwest Europe, the geography of the seas, and political with social geography.

a. *The maritime areas of Europe*

On the first theme, the maritime countries of northwest Europe, Vallaux's work was marked not only by a comprehensive view with a descriptive element (which he later stated to be fundamental premises of geography) but also by a desire to explain observed phenomena adequately, even if this meant seeking data from outside the apparent limits of geographical study. One such study was a chapter of his thesis on the system of land tenure in Brittany, which Vidal de la Blache, his supervisor, thought to be law rather than geography, as Vallaux observed in *La géographie de l'histoire*, 1921 (p.64), published after Vidal de la Blache's death in 1918.   An apparent digression into law was in fact necessary to explain the characteristics of the Breton landscape but such an apparent deviance from geography was unusual at the time. Then, as now, however landscape analysis included a scientific treatment of its structural characteristics.

   A conscientious -- indeed scrupulous -- worker, Vallaux was faithful to his scientific convictions in his zeal for field enquiry, made the more necessary by the inadequacy of statistical sources.   Such field studies were then new, and he used them in all his various travels to good effect.   But as the territorial range of his work widened, his fieldwork was necessarily less thorough.   Nevertheless the technique remained and had its fruits in the effective evocation of different landscapes.   His observation of the wind-swept interfluves of Jersey as a contrast with the luxuriance of the sheltered valley is enshrined in clear and decisive but at the same time colourful prose.   He thought that words could say more than statistics ('all numbers are dry and colourless, including means' *L'archipel de la Manche*, 44).   Thanks to its literary qualities, his geographical work was read by a wider circle than that of professional geographers, though its scientific accuracy was retained.   His book *Sur les côtes de Norvège* of 1923 was based on lectures he gave during a course organized by the *Revue scientifique* in 1912, with some earlier lectures.   The aim was to teach people willing to learn the art of observation and this, said E. Colin in the 1923 *Bibliographie géographique*, was splendidly achieved for the book had 'descriptions at one and the same time accurate and artistic ... yet raising fundamental geographical problems showing the fine scientific mind of its author'.

b. *Geographical oceanography*

As a contributor to oceanography Vallaux was both a teacher and a researcher.   The former task was fulfilled by writing a number of papers on the researches of others, as seen by a geographer.   Like all people whose interest is dual he faced the problem of apparent intellectual fission, for he was a geographer in oceanography and an oceanographer in geography.   However his dual interest was splendidly reflected in his large work of synthesis, *Géographie générale des mers* of 1933. Advertised as the epitome of all available knowledge, physical and human, about the ocean, it is his best known work. In fact the book is not so much a general as a regional survey for each ocean is treated separately. The idea of a regional geography of the seas was new in France though G. Schott, a German worker, had dealt specifically with the Atlantic.

c. Political and social geography

Although *Le sol et l'Etat*, 1911, is on much the same theme as F. Ratzel's *Politische géographie* the emphasis of Vallaux's work is sharply different for he placed the emphasis mainly on human rather than physical facts, designedly so as he was opposed to physical determinism. To him the term 'social geography' meant the geography of societies, the organization of human groups. This was a geographical factor equal in significance to the study of the natural environment, and at times more relevant. His aim was to remove geography from the domination of laws of the natural sciences, not to make it in any way less scientific, but rather to give it individuality of method, with the acquisition of the then nascent laws of sociology. In this view Vallaux once again shows his wish, apparent in his first thesis, to use the light shed by cognate disciplines on geographical phenomena. Having accepted this view he gave himself the task of assessing fully the methodology of geography.

B. Geography and Other Sciences

In *Les sciences géographiques*, Vallaux expressed his general view of geography and its place among other disciplines with a critical enquiry on the individuality of its methods. The use of the plural in the title of his book implied that there were several distinct aspects. Of these the first was an 'autonomous geography' and the second a group of 'auxiliary geographies', in effect those associated with sciences inevitably in association with space, for example botony, zoology, history.

Having said that 'geography has become scientific since it has provided explanations' and was therefore no longer merely descriptive, Vallaux saw its major purpose as the study of the interaction of varied phenomena on the surface of the earth. In this respect it was concerned with the world, with 'facts as a whole and their connections' while neighbouring sciences were 'molecular' as it was their privilege to analyse single aspects of reality. Geography borrowed from such sciences only to illuminate the connection of phenomena, but always it sought a real synthesis, at once complex and changing. While all the 'molecular' disciplines aimed at making a synthesis from detailed study -- 'Synthesis can only be made from analysis' Fustel de Coulanges advised -- geography on the other hand began with the whole world, with world forces and their interconnections.

And such study began with the landscape that could be seen within the horizon. But the study was not restricted to what was visible for there were other distributions -- such as density of population, migration or production -- which 'though not visible as a reality were still necessary to full understanding'.

Having accepted the idea of the reciprocal influence of phenomena and having seen their existence and juxtaposition, there is a logical progress to the regional synthesis, regarded as the major achievement of geography during the interwar years. But Vallaux was fully aware of its inadequacies: as geographical distributions had no simply definable limits, to divide the world into a kind of 'marquetry and mosaic' was to make the region 'a logical device and an educational method' at variance with reality. Therefore he refuted the idea that regional study was the crown of geography for to him it was an abstraction lacking real identity, especially as regions based on human and on physical features rarely coincided. Contrary to many other geographers, he held the view that 'the regional framework cannot unify the two aspects of geography, but in fact inevitably divides them' (*Les sciences géographiques*, 1925, 167). The solution lay in discerning large regions, terrestrial zones, more appropriate for showing major geographical correlations. He listed the world's zones with the great mountain ranges as fundamental zonal geography. While advocating this new geography he saw its limitations for human groups were rising above natural limitations. Human geography as 'a unifier and leveller' may supersede regional geography and zonal geography: indeed 'study of human geography shows that the regional map of the world is increasingly simplified' (*ibid.* 172).

This method is linked with the appreciation by geographers that some facts are permanent and unchanging. Two temptations must be avoided: of these the first is to make theories stronger in logic than in an appreciation of reality, such as the theory of destructive economies favoured by some economists; the second is the idea of the final, and discernible, cause of all phenomena and evolution, destructive of scientific advance. So conceived, geography 'having melted down in its crucible a multitude of facts from other sciences', may be able to provide a relative, though not an absolute, interpretation of man and his world. Even so, Vallaux insisted that he had no wish to be the founder of a science of sciences or a new metaphysical philosophy. Thanks to its own methods, geography had become a science that could bring much to other sciences as well as borrow from them. This attempt to provide a rational approach to geographical thought -- then having literary and even poetic qualities -- was new in 1925 and was based on Vallaux's philosophical views.

C. Geographical Thought

a. The rejection of determinism, though not of causality

Vallaux's endeavour to associate geography with the social sciences gave it a liberal quality antithetical to strict determinism, for 'man's presence in nature is the greatest of all obstacles to the generalisations of scientific determinism' (*Les sciences géographiques*, 1925, 180). In this view he followed the concept of 'possibilism' associated with Vidal de la Blache, and discarded the physical determinism of Taine and the racial determinism of Gobineau. States were not expressions of natural areas and only rarely were their frontiers natural features. In *La géographie de l'histoire*, 1921, written in collaboration with

Jean Brunhes, the central theme is an analysis of
the dual influence of human and natural features in
the evolution of states, 'the geography and the
history'.   The purpose of the authors was to detach
political geography from physical determinism.   But
to avoid misunderstandings, one must recognize that
Vallaux was conscious of causality and did not think
that events happened by mere chance;   in fact he held
firmly to scientific principles and thought that
society had its own laws which in time could be
discovered.

b. *Rationalism and realism in geography*

Vallaux remained faithful to his rationalistic atheism
and naturally from this came the views he held on the
place of man in nature and his faith in science.   To
him it was impossible for a scholar to accept any
religion which opposed science by speaking of 'God's
inscrutible purpose' and which made man 'the crown of
all living creation'.   Such an anthropocentric view
gave a false vision of the real world and Vallaux
eagerly accepted the atheism of the Copernican revolution when he wrote that 'For science ... human
conscience is just a fact and does not endow any man
with any exceptional dignity' (*Les sciences géographiques*, 277).   To have a conscience was not to
have a soul.   This atheism sustained his faith in
science and in its almost unlimited potentialities.
The supernatural element was a delusion for all could
be explained by rationalism and his ardent and tireless
purpose lay in seeking this explanation.

Having rejected anthropocentricism, he turned
naturally towards a philosophical realism.   Man was
an element in a reality which existed independently of
any individual, having in himself no power to create
existence as the idealists hold.   This view is frequently apparent in Vallaux's writings, notably in his
search for the laws which control the interaction of
phenomena and in his appreciation of concrete and un-
doubted facts.   These gave a view of reality against
which theories could be discarded, for our systems
of thought, even if logical, are merely tools we use
to understand reality and not a substitute for such
understanding.   He discarded all temptations of
abstraction, generalization and systematization for
the method of the geographer, always rigorous, was
essentially empirical.   Therefore to adopt an idea,
a concept such as that of the natural region, and to
attempt to superimpose it on the real world, was a
negation of the geographical approach which on the
contrary was rooted in a real world and discerned its
variety without preconceptions, for its foundation lay
in the unity of the landscape.

With such strongly defined philosophical views,
Vallaux had become a disciple more radical than the
Master of whom he wrote 'Vidal has never wished to
define his thought and give it precision...'.   It may
well be that Vallaux's forceful convictions and independence of spirit explain his scientific isolation
and with that the neglect of the substantial contribution to geographical writing during his life.

3. *INFLUENCE AND SPREAD OF IDEAS*

Despite the abundance of his publications, their
originality and variety, they are now little known.

This may be because he worked outside the main Faculties of the universities, but perhaps also because
his choice of subjects was unusual.

a. *A marginal geographer*

Vallaux never had the honour of holding a university
chair, but they were few in his day.   He spent much of
his life beyond easy contact with gatherings such as
conferences, congresses and excursions.   Though geographers such as J. Brunhes and E. Colin were personal
friends, his intellectual association was mainly with
scholars in other disciplines.   Nevertheless his
colleagues recognized the quality and depth of his
work.   A. Demangeon, in his comment in the *Bibliographie géographique* on *L'Archipel de la Manche* said
that it was 'a well constructed and well written book
which gave ... a living and personal description that
could not be recommended too strongly' despite the
'inadequacy of the maps' and the 'excess of illustrations of the coast with few from the rural areas'.
Geographers were at that time more interested in morphology or rural life, so Vallaux's work on the coastal
areas seemed marginal.   His work on the theory of
geography also attracted little general attention,
though there were exceptions.   In a non-commital review of *Les sciences géographiques* in *Annales de géographie* Ph. Arbos gave no clear reaction to Vallaux's
views and merely said that 'One might wish that these
questions (of methodology) were approached in a less
abstract fashion'.   Vallaux's views on regional
geography and on natural regions in particular appear
to have attracted little notice, though the idea of a
zonal perspective was to become favoured twenty years
later.

Fortunately colleagues had the wisdom to appre-
ciate the ability of Vallaux as an oceanographer.   He
attended the 1938 International Geographical Union
Congress at Amsterdam and gave a paper.   The *Géographie générale des mers* was well received and its
fine bibliography drew special praise though some
criticism was made of a lack of geographical balance,
because human geography, including that of the coasts,
received little attention.   'The inadequacies -- if
they are such -- are more those of the subject than
of the author' said Commander Marguet in a review
published in the *Annales de géographie*, 15 March 1934,
179.   He went on to say that the work gave a comprehensive treatment of physical oceanography, though it
was less convincing on the sea swell, waves and tides
as the author had not had 'the opportunity to study
mathematical and physical sciences adequately'.   Yet
he praised Vallaux in his conclusion for the book, and
especially for appreciating the significance of the
subject.   In effect it was a pioneer work but this
had not deterred its author for 'the best way to learn
to swim is to throw oneself in water'.

Vallaux attracted readers far beyond the normal
geographical public.   Through his articles in the
Revue générale des sciences and the *Mercure de France*
he reached many who appreciated his skill in synthesis
and clear expression.   His restatement and popular
presentation of geographical material gave him the
position of an ambassador for the subject to an
educated public.   At their time his polemics against
the Trans-Sahara railway and the theory of oceanic
transgressions of Le Danois were well known.

He was respected beyond France, and was made a corresponding member of the American Geographical Society. In 1918 and 1924 he wrote articles for the *Geographical Review*, which had long and eulogistic notes of his various works. In America he was regarded as one of the major representatives of the French geographical school.

b. The lack of continuity from his work
Vallaux's ideas have received little attention since his death. In his career he had no opportunity of teaching geographers. In his *Sciences géographiques* he distinguished 'auxilliary geography', and this was what he taught to his pupils at the Naval College and the Ecoles des Hautes Etudes Commerciales. In the lycées he could only hope that some pupils might be inspired to become geographers, and as he never had a University post he had no experience of directing research, and therefore no opportunity of making disciples. This is why his new ideas, such as zonal geography or the 'regional' geography of the seas, have not been followed. And it also explains why oceanography has attracted so few geographers; Vallaux as an oceanographer was on the margins of the university world of future teachers and researchers. This isolation has inevitably limited the influence of his work.

c. Vallaux's place in geography
His death in 1945, a time of troubled world conditions, was little noticed. A few brief obituaries appeared, noting some major stages in his career but without much comment on its scientific significance. A more adequate treatment appeared in the *Geographical Review* but it was mainly concerned with his connections with the journal. Nowadays his work is still little read and he is chiefly remembered as a specialist in oceanography whose work is understood to be out of date. The originality of his ideas and his views on methodology are ignored. But this injustice has, as noted earlier, come from his long isolation in Brittany, outside the universities. As fate decreed that this scholar from the Ecole Normale Supérieure should not go with others into University work, he had to find a harder and humbler way which fortunately did not inhibit a career in research but imposed other restrictions in his teaching and influence. However now that methodological studies are once more in favour, his work in the field has been rediscovered. Consequently, recent studies on the history of geography have given him more attention than he received in his lifetime, for he is regarded as one of the most far-seeing of the disciples of Vidal de la Blache, whose thought he has explained with vigour and clarity. And it should not be forgotten that he was the pioneer of geographical oceanography in France.

Bibliography and Sources

1. OBITUARIES AND REFERENCES ON CAMILLE VALLAUX
No complete bibliography exists, but during this study more than 140 articles and books were listed, of which a selection is given here. Nor has there been any published article on Vallaux, apart from three brief obituaries:
Colin, E. 'Camille Vallaux', *Ann. Géogr.*, vol 54 (1945), 305
Parret, R. 'Camille Vallaux', *Bull. Assoc. Géogr. Fr.*, no 175 (1946), 47
Anonymous 'C. Vallaux', *Geogr. Rev.*, vol 36 (1946), 164

2. SELECTIVE AND THEMATIC BIBLIOGRAPHY OF WORKS BY VALLAUX

a. Countries of north west Europe
1907 'La Basse Bretagne, étude de géographie humaine' (Thesis), Paris, 320p.
1907 'Penmarch aux XVIème et XVIIème siècles' (Thesis), Paris, 43p.
1913 *La Norvège, la nature et l'homme*, Paris, 115p.
1913 *L'archipel de la Manche*, Paris 256p.
1923 *Sur les côtes de Norvège*, Paris 189p.
1924 'The maritime and rural life of Norway', *Geogr. Rev.*, vol 14, 505-18
1925 *Visages de la Bretagne*, Paris (in collaboration)

b. Oceanography and geography of the seas
1908 *Géographie sociale: la mer*, Paris, 377p.
1932 *Mers et océans*, Paris, 100p.
1933 *Géographie générale des mers*, Paris, 796p.
1938 'La circulation de profondeur dans les grands océans', *C.R. Congr. Int. Géogr.*, Amsterdam, Tome 2, Sect. 2, 27-44
1942 'Recherches récents sur la circulation superficielle et profonde de l'Atlantique Nord', *Ann. de Géogr.*, vol 51, 161-74

c. Political geography
1911 *Géographie sociale: le sol et l'Etat*, Paris, 420p.
1918 'German colonization in Eastern Europe', *Geogr. Rev.*, vol 6, 165-80 (with Jean Brunhes)
1921 *La géographie de l'histoire*, Paris, 216p. (with Jean Brunhes)

d. Methodology and epistemology of geography
1925 *Les sciences géographiques*, Paris, 407p.
1927 'Les méthodes d'observation en géographie', *Revue de métaphysique et de morale*, vol 34, 473-88

e. Biographical studies
1926 'L'oeuvre scientifique d'Alexandre de Humboldt', *Rev. Sci.*, vol 64, 491-8
1930 'Jean Brunhes', *La Géogr.*, vol 54, 237-9
1936 'Thoulet (J.O.) 1843-1936', *Ann. de Géogr.*, vol 45, 217-18
1946 'Jules Richard, l'homme, le savant, l'organisateur', *Bull. Inst. Océanogr.*, Monaco, no 892, 11-15

f. Varied works
1899 *Les campagnes des armées françaises (1792-1815)*, Paris, 368p.
1919 'Péronne', *La vie urbaine*, vol 1, 77-95
1920 *Etude économique sur le projet du canal de la Loire à la Manche par la Sarthe ou par la Mayenne*, Paris, 71p.

1922 'Montdidier', *La vie urbaine*, vol 4, no 14, 20p.
1922 'Un petit pays de la vieille France: le Ven-
     dômois', *La Géogr.*, vol 36, 165-79
1923 'Le bassin parisien', *Géogr. Univ. Quillet*, Paris,
     vol 1, *Le monde français*, 269-347
1924 'Les projets de chemin de fer à travers le
     Sahara', *Mercure de France*, Paris, 1 March,
     309-30
1927 'Le transsaharien', *Bull. Assoc. Géogr. Fr.*,
     vol 36, 563-76
1929 'L'erreur transsaharienne', *Rev. Sci.*, vol 67,
     69-75

*François Edmond Marcel Michel Carré, Agrégé de géo-
graphie, is maître-assistant de géographie at the
University of Paris-Sorbonne.   Translated by
T.W. Freeman.*

Dates	Life and career	Activities, travel, fieldwork	Publications	Contemporary events and publications
1870	Born at Vendôme (Loir-et-Cher)			Thesis of Boutroux, *De la contingence des lois de la nature*
1891	Entered Ecole Normale Supérieure	Fellow students: Ed Herriot, M. Zimmermann, friend of J. Brunhes, pupil of Vidal de la Blache		Thesis of Jaures, *De la realité du monde sensible*
1894	Agrégation d'histoire et de géographie			Ratzel, *Politische géographie*
1895-6	Taught at Lycée, Pontivy			
1897	Taught at Lycée, Brest			
1899-1901			First book: *Les campagnes des armées françaises (1792-1815)*	
1903	Professor of Geography, Ecole Navale, Brest	Visit to Channel Islands		Vidal de la Blache: *Tableau de la géographie de la France*
1907		Cruises with Ecole Navale, Scotland, Norway, Baltic	Doctorate theses: 1) *La Basse Bretagne* 2) *Penmarch aux XVI ème et XVIIIème siècles*	Bergson, *Creative Evolution*
1908	Settled at Kerhuon			
1910		Mission to Jersey for Ministère d'Instruction Publique	*Géographie sociale: la mer*	
1911		Visit to Channel Islands	*Géographie sociale: le sol et l'Etat*	
1912		Lectures on Norway		
1913	Professor at the Lycée Buffon (Paris) lectured on Brittany at Ecole des Hautes Etudes Sociales	Collaboration with Jean Brunhes	Cruises described in *La Norvège, la nature et l'homme* and in *L'archipel de la Manche*	Schott, *Géographie des Atlantischen Ozeans*
1914				First World War
1919	Taught history at the Lycée and geography at the Ecole des Hautes Etudes Commerciales (Paris)	President of the Geography Commission, Ligne maritime et coloniales: tour of North Africa		
1921	Examiner at Ecole Navale	Wrote the oceanographical reports in *La Géographie*	*La géographie de l'histoire* (with J. Brunhes)	
1922		Corresponding member, American Geographical Society		Febvre, *La terre et l'évolution humaine*

Dates	Life and career	Activities, travel, fieldwork	Publications	Contemporary events and publications
1923			*Sur les côtes de Norvège* *La Bassin parisien*	Le Danois, *Etude hydrologique de l'Atlantique Nord*
1925			*Les sciences géographiques*	
1928		Compiled oceanographical reports for *Ann. Géogr.*		
1932	Retirement: lived at Paris and Kerhuon		*Mers et océans*	Le Danois, *Etude hydrologique de l'Atlantique Nord*
1933			*Géographie générale des mers:* wrote numerous articles in *Annales de géographie* (last in 1942), *La géographie, Bulletin de l'Institut Océanographique, Revue générale des sciences, Revue scientifique*	Le Danois, *Les transgressions océaniques*
1938		Gave paper on marine hydrology at I.G.U. Congress, Amsterdam		
1939		Visit to Monaco (International Hydrographic Laboratory)		Second World War
1945	Died at Kerhuon		*Visages de la Bretagne* (in collaboration)	

George Vâlsan
1885-1935

NICOLAE POPP

Cruelly injured in an accident early in his career, Vâlsan managed to establish geomorphology firmly in Romania as a basis for the study of the *paysage*, the rural landscape. He wrote papers on historical and social geography, on methodology and the theory of geography, and on geomorphology. Having studied in Germany and France, he turned more to France for inspiration, and especially to Emmanuel de Martonne. His writing combined scientific strength with fine expression and revealed a cultivated mind.

1. EDUCATION, LIFE AND WORK
Perhaps the ablest and certainly the most courageous of all Romanian geographers, Vâlsan was born on 24 January 1885 into a family of railway workers. He entered Bucharest University in 1904 as a student of philosophy under the logician Titu Maiorescu and the geographer Simion Mehedinţi. Appointed as an assistant lecturer in 1908, he published his first paper, on the ground underlying the town of Bucharest, in 1910, and in the following year he went to Berlin with a scholarship from the Romanian Geographical Society, to study geography with Albrecht Penck and ethnography with Felix von Luschan. In 1913 he went to Paris as a student of Emmanuel de Martonne, with whom he formed a warm friendship. De Martonne expected that Vâlsan would do much to advance modern geography in Romania and his doctorate thesis on the Romanian Plain, 1915, was a notable contribution to its literature.

Appointed a professor at Iasi, Vâlsan was unable to do much research during the war years, and in 1917, on his way to the front line, he was severely injured in a railway accident that left him partially disabled for the rest of his life. In 1919 he moved to Cluj University, where he built up a strong school of geography and also founded a journal, the *Travaux de l'Institut de Géographie, Cluj*. In 1920, at the early age of thirty-five he became a member of the Romanian Academy and in 1929 he moved to the university of Bucharest where he held the chair of physical geography until his premature death on 6 August 1935. Having excellent French and German, he was a valued representative of Romania at International Geographical Union congresses in Cambridge 1928, Paris 1931 and Warsaw 1934. At the Warsaw meeting he was a vice-president of the section on *Géographie de paysage*, the rural landscape.

Convinced that geography teaching was of great value in schools, Vâlsan gave much thought to its teaching. He was one of comparatively few professors who taught students in their first year and his advice to them was:

> Accept with courage the learning of science for yourselves, attack its problems as a personal commitment, and verify your findings, whenever you can, by contact with nature.

To him learning was life and specialism a challenge. Geography was part of a wider field of science.

2. SCIENTIFIC IDEAS AND GEOGRAPHICAL THOUGHT
Physical geography in general and geomorphology in particular were Vâlsan's main interests but he never neglected the social aspects of his subject. His

detailed researches on physical geography led in time
to works of synthesis on the Danube plain and delta,
and he also wrote on problems in historical geography,
such as a study of the former population of Wallachia.
In his work he strove for perfection and made several
drafts of every paper. His poor health compelled him
to abandon some of his projected publications of which
a number were published posthumously.

a. *Physical geography*

His fundamental work, preceded by the publication of
ten papers from 1909-14, was his thesis on the physical
geography of the Romanian Plain. This work had been
begun under the care of Emmanuel de Martonne, to whom
Vâlsan gives a generous tribute in his preface. All
studies in geomorphology and many also in geology
written since Vâlsan's thesis appeared in 1915 have
recognised its merit for they have confirmed his conclusions and found his intuitive suggestions valid.
He was the first worker to demonstrate that on the
margins of the sub-Carpathian upland the Plain was
dominated by terraces, but in a central area there were
no terraces at all while they occurred in the area in-
undated at various times by tributaries of the river
Danube. Vâlsan decided to call the area the Romanian
rather than the Danube Plain, as in his view it was not
created by the river at all. Rather an explanation
of its form had to be sought by studying surfaces of
erosion, notably from the years 1919-25 in the upper
reaches of the Prahova river where he traced the
evolution of the physical landscape from the later
Pliocene era to the present time.

Earth movements in the Plain were obviously of
crucial importance, and this led Vâlsan to study the
delta of the Danube and the area of the Iron Gates
which, he maintained, was a defile formed by river
capture as E. de Martonne suggested and not by super-
imposition, the view of Jovan Cvijić. Therefore the
Danube valley, in the Iron Gates area, was not a defile
separating the Carpathians from the Balkans but a
transverse valley within the Carpathians. From this
it followed that the mountains between the Timis-Cerna
corridor and the Timoc valley were one entity, which
he named the 'Iron Gates massif', consisting of a
recently elevated peneplain, in which the Danube and
its tributaries had cut narrow valleys with dramatic
gorges, having calcareous rocks both to the west and
east. This massif, he argued, separated the Romanian
from the Pannonian Plain.

From 1926 to 1934 Vâlsan's interests turned again
to the mouths of the Danube and the coasts of the Black
Sea. In his paper given at the Warsaw Congress of
1934 he advanced the bold hypothesis that the Danube
delta resulted from regression rather than advance of
the land. In his view the delta had been formed quite
recently, built up in a marine gulf formed after the
loess had been deposited on a steppe landscape with
lakes. He adduced various types of evidence for this
view, which involved such considerations as changes of
level due to eustatic movements in the late glacial and
post glacial epochs.

Clearly Vâlsan was a worker permanently conscious
of the need for detailed fieldwork in his researches.
At the same time he was searching for a general method
of geomorphological research and expressed this in
papers on the evolution of valleys (1918), the spatial

element in geographical description (1928) and the
fundamental processes by which the earth's crust is
moulded (1932). He found it surprising that no manual
of geography had a chapter dealing with spatial reality
and its elements, for though no one doubted the value
of a map, nor the importance of descriptive and explanatory geography, this was the only discipline that
could successfully combat criticism put forward by
detractors of geography. It was in this lack of a
study of spatial reality and its elements that one
could discern the immaturity of scientific geography.

By his work in the Romanian Plain, in the
Carpathians and the Danube delta with the Black Sea
coasts, Vâlsan abundantly demonstrated his power of
analysing and explaining the spatial reality of the
physical landscape. And this work was the basis of
his geographical theory and methodology, in which five
main principles emerge. The first was that in education one proceeds from the known and local to the
unknown and distant, with a deepening understanding of
the home neighbourhood. Secondly, he was a pioneer
in the study of the landscape, the *paysage*, which to
him was the only geography having methods and laws
distinctive to itself, shared by no other natural
science. Thirdly, space, definable and measurable,
was itself an element fundamentally and essentially
geographical. Fourthly, he showed the significance
of the microgeography and microphenomenology of nature
as essential components of an alphabet for the study
of physical geography. Finally, Vâlsan was the first
Romanian geographer to bring out the importance of
soils (1930) and so he is known as the pioneer soil
geographer of his country.

b. *Human, historical, ethnographic and methodological geography*

George Vâlsan was a complete geographer. He recognized the existence of several geographical sciences
but thought that before anyone became a specialist in
any physical, social or any other aspect of the subject
he should know the techniques of investigation and
possess the mode of thought characterizing geography as
a whole. De Martonne was to him a living example of
this view: indeed all the great geographers of his day
were such people. No specialism should be followed to
a point where the geographer as such would detach it
from the mainstream of geography and so change his view
of geographical thought.

Holding these views, Vâlsan confirmed them by
writing valuable work on aspects of geography other
than geomorphology. In human and historical
geography he was always concerned with the relation
between societies and the geographical setting, between
man and environment. This was seen from 1912 when,
having found in Berlin a Russian statistical map of the
nineteenth century, he wrote an article showing that
before 1850 the sub-Carpathian region was more densely
populated than the Plain. This idea, maintained
through several years of varied research was developed
further by 1928 when he published a deeply thoughtful
study on the physical environment and the national
biological assets.

In research on the relationship between the
physical environment and a people one meets
an immediate problem. Logically one must
accept that the physical environment must

influence a people. This is a general conviction, a postulate of all science which
has investigated the relation between man
and his setting, whether one begins with the
environment as in geography or with man as
in sociology, enthnography or history. An
ethnic organism is malleable and so imprinted
by environmental influences. But organisms
are not passive for in their life they are
ceaselessly creating their environment and
therefore moulding their surroundings.
Therefore between a people and their environment there is an un-intermitting interaction
so that the evolution of a people is not
passively imposed by the environment but is
rather an active and continuous adaptation,
a compromise between environmental guide-
lines and the response given by any ethnic
organism.

To deal with these problems is not easy for as
Vâlsan said 'It is much easier to assess the relationship between the environment and the density of population than to discern in the ethos of a people what
makes them respond to physical factors such as relief,
climate and waters'. Therefore he recommends that the
researcher 'should proceed with great prudence if he
wishes to remain within the limits of scientific enquiry'. In Romania Vâlsan found a dual 'personality'
which had been immensely significant in the history of
its people -- the Carpathians and the Danube. These
should not only be regarded as separate and individual,
but also in co-operation, for there was integration
between the pastoral and agriculture life of the
Carpathians with their adjacent mountains and the commercial and fishing wealth of the Danube.

In the 1920s and 1930s geographers did not relate
the present to the future, still less the past to the
future. Generally human geography was the science of
the actual human landscape and historical geography was
the human geography of the past. Vâlsan showed time
and again that the relation between societies and environments was incessantly changing. This implied
that in the future both societies and the environment
in which they lived would be vastly changed.

While working at the St. Geneviève Library in
Paris in 1914 Vâlsan found a map of Moldavia prepared
by Prince Démètre Cantemir (1673-1723) which he
considered to be a major cartographic discovery.
It completed his study of the work of Cantemir,
whose *Descriptio Moldaviae*, which he had edited in
about 1910, was 'the best geography of Moldavia ever
written!' In his three articles on Démètre Cantemir,
Vâlsan spoke of this prince of Moldavia (who was also
a member of the Academy of Sciences in Berlin and
councillor to Tsar Peter I) as the forerunner of modern
Romanian geography. The map found in Paris was valuable not only for its distinguished cartography but
even more for its content as it gave a precise and
complete image of Moldavia at the beginning of the
eighteenth century.

Among his many scientific interests Vâlsan was
particularly attracted to ethnography. When studying
at the Humboldt university in 1911 he gave to the
Romanian Academic Society of Berlin a talk on 'Our
ethnographical obligations'. His interest was seen in
his 1927 paper on ethnography as a new science and he
was eager to bring ethnography and geography together
as they were both field studies. He commented that
'Ethnography is more a science of observation and
analysis in a natural environment than a study of
texts' and that it could become a 'true science'.
Geography and ethnography, jointly with history and
philology were the only four fundamental sciences
needed to study the specific character of a country and
of the people who lived within it.

Vâlsan was particularly concerned with the writing
of monographs of small regions (*pays*), considered
geographically, including their ethnic, social and economic characteristics. This led him into sociology,
but in his 1929 paper on geography and sociological
researches he criticized sociologists for writing
studies of individual towns and villages on the
ground that no town or village was an isolated entity
but part of a wide human world. Only a regional study
could be effective, the more so if it took account of
the complexity of geographical influences, some of them
historical, especially if such a study was the work of
several people closely associated with the area under
scrutiny. There was no single influence dominating
all others and Vâlsan, though himself a distinguished
specialist in geomorphology, was critical of geographers who regarded physical qualities as the key to
explanation of the geographical characteristics of any
area. This was finely stated in these words:
> We do not deny that the geographical element
> is crucial, yet we must not support geogra-
> phical determinism for the physical qualities
> of the earth are actual or potential facts
> which acquire human value and significance
> in so far as people have the capacity for
> creation, by their work, persistence and
> intelligence.

3. INFLUENCE AND SPREAD OF IDEAS

In his works published during his short life and posthumously, of which only his thesis on the Romanian
Plain was lengthy, Vâlsan dealt with all aspects of
geography and also with those sciences closely connected with it. His work, marked by close reasoning
and concise writing, gives an admirably clear view
of his findings. Though a specialist in physical
geography, he never neglected the people who moulded
the landscape.

He was at one and the same time a scholar devoted
to his researches and the professor devoted to the
education of his students. His researches on the
Romanian Plain (1915) and the coasts of the Black Sea
(1934) remain basic for all who study these areas, and
with these must be mentioned his work on surfaces of
erosion and on fluvial terraces. He was permanently
conscious of the spatial element in all geographical
description and was one of the first geographers in
the world to conceive geography as a science of the
paysage. For Romania he showed, as nobody did before
or after his time, the role of the Plain and the
Carpathians in its development, and E. Bacuta noted
that in him there was a poet eager to show the beauty
of his homeland. His theory of geography involved
not only his emphasis on the *paysage*, but also on the
local scene.

In 1931 he acquired fame as a reformer in Romanian

geographical teaching, by his campaign to replace the
current practice of proceeding from 'the general to the
particular' by working 'from the near to the far, from
the known to the unknown'.   This has now been generally accepted by educationalists, but from a principle
existing only in embryo, Vâlsan worked forward to what
in time became the study of the environment.

The only honour given to Vâlsan was his membership
of the Academy, but he was known as a dedicated scientist both nationally and internationally.   He was the
creator of the geography school at Cluj University,
where his tenure of the Chair brought lustre to
Romanian geography.   Many students were trained for
doctors' degrees, and from 1922 he directed the publication of the *Travaux de l'Institut de Géographie de
l'Université de Cluj* and also the *Bulletin* of the
Romanian Geographical Society, in his day the most
important geographical periodicals of Romania.   His
courses at Bucharest University, 1930-5, on geomorphology, mathematical geography, cartography and
biogeography, were models of academic presentation and
scientific content.

Internationally he was known for his membership of
the International Geographical Union Commission on
Pliocene and Pleistocene terraces, on which he made
contributions on Romania to the international congresses at Cambridge in 1928, Paris in 1931 and Warsaw
in 1934.   He was a vice-president of the Physical
Geography section at Paris and the *paysage* section at
Warsaw.   Altogether, with those published after his
death, a full bibliography of Vâlsan includes 182
titles, marked by both scientific content and literary
quality.

His time as a student in Paris, his friendship
with Emmanuel de Martonne and his affection for France
made Vâlsan a representative of French culture especially attached to the views of the French school of
geography.   Robert Ficheux spoke of Vâlsan as the most
distinguished of Romanian disciples of French geography
and notes that towards de Martonne he had a somewhat
diffident air of awe and respect, while de Martonne had
for his young disciple, so cruelly impeded from travelling by infirmity, a quiet admiration and a discerning
affection.   Both de Martonne and Vâlsan were marked by
intellectual integrity and in each the man of science
was combined with the soul of an artist.   But while
de Martonne was forceful, energetic, confident, enthusiastic and optimistic, Vâlsan was always febrile,
uncertain of the future, though eager to give his
country the best he could offer in his poetic reveries,
mature reflections and finely constructed geographical
works.

A life of suffering led him to the philosophy expressed in the concise statement that 'An individual is
only the temporary bearer of an eternal life which is
handed on through the centuries'.   Vintilă Mihăilescu,
his successor in the Chair of geography at Bucharest
University, wrote that 'With astonishing courage, he
was strengthened by this somewhat ironic philosophy and
remained faithful to it, although the tragedy of his
life meant that he had to make his contribution to
learning in times of respite from constant and devasta-
ting illness'.   On his death the famous historian,
Nicolas Iorga, spoke with admiration of Vâlsan as one
who had given geography new light and also showed in
his work that scientific accuracy could be combined
with eloquence of writing.   Vintilă Mihăilescu, in his
panegyric, spoke of his courage, his heroism and his
contribution to the culture of his country.   And his
influence went further, for he was known far beyond
Romania for his distinguished work.

Bibliography and Sources

1. OBITUARIES AND REFERENCES ON GEORGE VALSAN

Mihăilescu, V. 'G. Vâlsan la geograf si educator
    (George Vâlsan, geographer and pedagogue)', *Bul.
    Soc. Rom. Geogr.*, Bucuresti, vol 54 (1935), 1-18
Milhăilescu, V. with Morariu, T., 'Bio-bibliografia
    lui George Vâlsan (a biobibliography of George
    Vâlsan)', with French abstract, *Bibl. Bibliogr.*,
    no 14, Cluj (1937), 53p.
Popp, N. 'George Vâlsan -- geograf international,
    spicuiri (George Vâlsan, international geographer,
    some observations)', *Rev. Geogr.*, vol 1/4 (1945),
    131-7
Morariu, T. and Mihailescu, V. 'G. Vâlsan -- Descrieri
    geografice (G. Vâlsan as a descriptive geograph-
    er)', in *Viata si opera lui George Vâlsan (Life
    and work of George Vâlsan)*, Bucuresti (1964), 5-50
*George Vâlsan -- Opere Alese (George Vâlsan --
    Selected Works)*, (Bucuresti, 1971), includes:
Morariu, T. 'Omul si opera (The man and his work)',
    11-37
Popp, N. 'Principalele probleme de geografie fizica în
    opera lui George Vâlsan (The main problems of
    physical geography in the writings of George
    Vâlsan)', 38-69
Conea, I. 'George Vâlsan, geograf al raporturilor
    dintre societate si mediul geografic in general
    si dintre poporul român si pămîntul românesc în
    special (George Vâlsan, geographer of the relation
    between society and environment in general and
    between the Romanian people and their country in
    particular)', 70-90
Onisor, T. 'George Vâlsan si etnografia românească
    (George Vâlsan and Romanian ethnography)', 91-128
Savu, A. 'George Vâlsan, pictor în cuvinte al pamîn-
    tului tarii noastre (George Vâlsan, an artist in
    his writings on Romania)', 129-48

2. BIBLIOGRAPHY OF WORKS BY GEORGE VALSAN

a. Physical geography
1910 'Bucurestii din punct de vedere geografic.
    Temelia Bucurestilor, (Bucharest: a geographical
    study.   The physical site of the city)', *Ann.
    Geogr. si Anthropogeogr.*, vol 1, 105-59
1913 'Observatii asupra cîmpiei romane orientale.
    Remarques sur les terrasses de la plaine roumaines
    orientale (Comments on the terrace of the eastern
    Romanian plain)', *C.R. Acad. Sci.*, Bucharest,
    vol 157, 185-6
1914 'Asupra morfologiei Olteniei.   Sur la morphologie
    de l'Olténie (On the morphology of Oltenia,
    Romania)', *C.R. Etud. Géogr.*, Paris, vol 157,
    1555-7

1915 'Câmpia Româna.   Contributii de geografie fizica
(The Romanian Plain: contributions on physical
geography)', with abstract in French, *Bul. Soc.
Rom. Géogr.*, vol 36, 313-568
1919 'Asupra trecerii Dunarei prin Portile de Fier
(The Danube passage of the Iron Gate)', with
abstract in French, *Bul. Soc. Rom. Géogr.*, vol 37,
133-52, 562-4
1918 'Vaile, originea si evolutia lor (The origin and
evolution of valleys)', *Bul. Soc. Rom. Géogr.*,
vol 37, 333-53
1925 'Alpii (The Alps)', *Rev. Adamachi*, vol 2, Iasi,
87-94
1928 'Les terrasses de la Plaine Roumanie (The terraces
of the Romanian Plain)', *Congr. Int. Géogr.*,
Cambridge, *Rap. Comm. Terrasses*, I.G.U., Madrid,
6-10
1934 'Nouvelle hypothèse sur la Delta du Danube (A new
hypothesis on the Danube delta)', *Congr. Int.
Géogr.*, Warsaw, vol 2, 342-53
1935 'Rapport sur les dernieres études concernant les
terrasses de rivieres en Roumanie (Report on the
recent studies of Romanian river terraces)',
Congr. Int. Géogr., Paris 1935, *Bul. Soc. Rom.
Géogr.*, vol 54, 19-22
1935 'Sur une plate-forme littorale en Roumanie (On a
coastal platform in Romania)', *Bul. Soc. Rom.
Géogr.*, vol 54, 23-31
1939 'Morfologia vaii superioare a Prahovei si a
regiunilor vecine (Morphology of the upper Prahova
valley and the surrounding regions)', *Bul. Soc.
Rom. Géogr.*, vol 58, 1-44
(The last three publications posthumous.)

b. Human and historical geography, ethnography
1912 'O faza în popularea Tarilor Românesti.   Cu
prilejul unei harti statistice vechi descoperite
in ultimul timp (A stage in the settlement of the
Romanian principalities, based on an ancient and
recently discovered statistical map)', *Bul. Soc.
Rom. Géogr.*, vol 35, 201-26
1920 'La terre et le peuple roumains (Romania: the
land and the people)', *Rev. Gén. Sci.*, Paris,
vol 31, 3-20
1925 'Opera geografica a Principelui Dimitrie Cantemir,
1673-1723 (The geographical works of Prince
Dimitrie Cantemir)', *Trav. Inst. Géogr. Univ.
Cluj*, vol 2, 3-20
1927 *O stinta noua: ethnografia (Ethnography: a new
science)*, Cluj, 43p.
1928 'Mediul fizic extern si capitalul biologic
national (The external physical environment and
the national biological stock)', *Bul. Eugenie*,
Cluj, vol 2, 1-26
1929 'Transilvania în cadrul unitar al pamîntului si
poporului românesc (Transylvania, part of the
unity of the Romanian land and people)',
Transilvania, Banat, Crisana, Maramures,
Bucuresti, 145-56
1935 'Dunarea (The Danube)', *Bul. Soc. Rom. Géogr.*,
vol 54, 38-55
1935 'Dobrogea (La Dobrogea)', *Bul. Soc. Rom. Géogr.*,
vol 54, 56-67
1937 'Orasele franceze în Evul Mediu (French towns in
the Middle Ages)', *Bul. Soc. Rom. Géogr.*, vol 55,
21-41

1937 'Evolutia Statului Roman in cadrul sau geografic
(The evolution of Romania in its geographical
framework)', *Bul. Soc. Rom. Géogr.*, vol 56, 20-37
1938 'Distributiunea în Romania a trei animale
disparute: bour, zimbru, breb (The distribution
of three extinct animals, the aurochs, European
bison and beaver, in Romania), with abstract in
French', *Bul. Soc. Rom. Géogr.*, vol 55, 22-36
1938 'Parisul, capitala lumii (Paris, capital of the
world)', *Bul. Soc. Rom. Géogr.*, vol 57, 37-47
(The last five entries are of posthumous works)

c. Other works
1924 *Povestea unei tinereti (The story of my Youth)*,
Bucuresti, 222p. (prose)
1925 *Gradina — parasita-poezii (The forsaken garden)*
(poetry), Cluj, 103p.
1929 'Elementul spatial în descrierea geografica
(The spatial element in geographical description)',
Trav. Inst. Géogr. Univ. Cluj, vol 4, 439-60
1936 'Cercetarile sociologice privite din punct de
vedere geografic (Sociological research: a geographical view)', *Bul. Soc. Rom. Géogr.*, vol 55,
1-20
1939 'Sensul geografiei moderne (The meaning of modern
geography)', *Bul. Soc. Rom. Géogr.*, vol 57, 1-21
1940 'Pamîntul românescu si frumusetile lui (The beauty
of the Romanian countryside)', Bucuresti, 169p.
(The last three works are posthumous)

*Nicolae Popp is Professor of Geography at the
University of Bucharest.   Translated from French by
T.W. Freeman.*

CHRONOLOGICAL TABLE: GEORGE VALSAN

Dates	Life and career	Activities, travel, fieldwork	Publications	Contemporary events and publications
1885	Born at Bucharest			
1904	Entered Faculty of Philosophy, University of Bucharest	Fellow student of C. Bratescu: pupil of S. Mehedinţi and T. Maiorescu		Publication of E. de Martonne, *La Valachie*
1908	Assistant at University of Bucharest			
1909				W.M. Davis, *Geographical essays*
1910			First article, on Bucharest	
1911		Student at Berlin University under Albrecht Penck and Felix von Luschan		E. de Martonne, *Les Alpes de Transylvanie*
1912			Article on 'Historical geography, based on an ancient map'	
1913		Student at Sorbonne, Paris, as a student of E. de Martonne		
1914				Outbreak of war
1915			Doctorate thesis on the Romanian Plain	
1916	Professor of Geography, University of Iasi			
1918			'The Danube passage of the Iron Gate'	
1919	Professor of Geography, Cluj University			
1920		Became member of the Romanian Academy Initiated field tours of the Institute of Geography, Cluj University		Visit to Cluj by E. de Martonne
1925	Became Director of the Institute of Geography, Cluj University		'The geographical works of Prince Dimitrie Cantemir'	W. Penck, *Die morphologische Analyse*
1928	Member of the I.G.U. Commission on Fluvial and Marine Terraces	Friend of N. Iorga, I. Cantacuzino and G. Opresco	'The terraces of the Romanian Plain'	I.G.U. Congress, London and Cambridge
1929	Professor of Physical Geography, Bucharest University		'The spatial element in geographical description'	

Dates	Life and career	Activities, travel, fieldwork	Publications	Contemporary events and publications
1931		Vice-President, Physical Geography section, I.G.U. Congress		I.G.U. Congress, Paris
1934		Vice-President, *Paysage* section I.G.U. Congress	'A new hypothesis on the Danube delta'	I.G.U. Congress, Warsaw
1935	Died at Agigea (Constanta)		Several works published posthumously	

Alexander Ivanovitch Voyeikov
1842–1916

I. A. FEDOSSEYEV

In the history of geography Alexander Ivanovitch
Voyeikov won renown for his outstanding works on
climatology and physical geography. He considered
climate to be an organic component of man's geographical environment, and concentrated on the changes
occurring in nature and the exploitation of natural
resources.

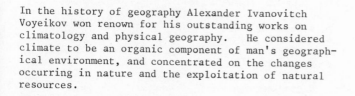

1. LIFE, EDUCATION AND WORK

Alexander Voyeikov was born in Moscow on 8 May 1842.
His father was Ivan Fyodorovitch Voyeikov, an officer
who had taken part in the 1812 campaign, retired in
1815, and from then on until his death, more than
thirty years later, spent his time farming his
estates. Both his father and his mother, Varvara
Dmitriyevna Mertvago, died when their son was five
years old, and A.I. Voyeikov spent his childhood on
the country estate of his uncle, D.D. Mertvago, near
Moscow. There he received an excellent education at
home and mastered English, French and German. In 1860
Voyeikov entered the Physics and Mathematics Department
at St. Petersburg University. When the University
was closed by the government due to student unrest,
he continued his education in Germany at Heidelberg,
Berlin and Göttingen universities. At Göttingen and
Berlin he studied meteorology. In 1865 he presented
his thesis at Göttingen University on 'Direct insolation in various parts of the globe' (Uber die directe
Insolation und Strahlung an Verschiedenen Orten der
Erdoberflache) and became a Doctor of Philosophy.
When he returned to Russia in 1866 he was elected a
full member of the Russian Geographical Society which
he served devotedly for the next fifty years. Established in 1845, this society was always concerned with
the study of Russia, because even as late as the
second half of the nineteenth century, much of its
area was little known. The northern parts of the
empire, including the Arctic coast and the plains and
mountain ranges in the east, were all awaiting exploration and Voyeikov worked on a number of committees
organizing scientific expeditions. While participating in the preparation of an expedition to study the
northern regions of Russia in 1870 he particularly
stressed the need to study these regions from a
meteorological point of view and correctly pointed out
that atmospheric processes at high latitudes might
have a marked effect on the climate and meteorological
conditions at medium latitudes.

After taking hard (as family legend has it) the
refusal of Sofia Kovalevskaya, subsequently a well-
known mathematician, to become his wife in the spring
of 1872 Voyeikov made no more attempts to marry and
devoted himself entirely to science and travelling
which he did with the money bequeathed by his parents.
He was extremely hardworking, enjoyed good health and
was a vegetarian. When still a youth he toured
western Europe, Palestine and Syria with his relatives.
In 1868 he travelled around the Caucasus and he went
there again a year later. In 1896 on the instructions
of the Geographical Society, he studied the organization of the meteorological service in western Europe.
This led to valuable contacts between the Society and
scientific institutions in western Europe and Voyeikov
became closely acquainted with many eminent meteor-
ologists, in particular J. Hann.

Still the travelling continued.   In the spring of 1872 Voyeikov toured Galicia, Bukovina, Wallachia, Transylvania and Austria to study the black earth soils; and in the autumn of the same year he set off again.   After visiting Vienna, Berlin, Gotha, Utrecht and London, he went to New York at the beginning of 1873.   From April till October 1873 he travelled around the USA and Canada; he visited New Orleans, Texas, Colorado, the Rocky Mountains, the Great Lakes and the Eastern States.   He then worked in Washington for three months at the invitation of the secretary of the Smithsonian Institution, Joseph Henry, completing the manuscript of J.H. Coffin's *Winds of the globe*, (Washington City 1875), to which he added data on the winds in Russia.   Almost the whole of 1874 and the beginning of the next year was spent travelling around tropical America.   Having visited Yucatan and Mexico, from Mexico City Voyeikov rode more than 1,000 kilometres through Tehuantepec, Tonala, Socanusco and across the Guatemala plateau to the town of Guatemala. From there he travelled to Panama by steamer and then around South America to the river Amazon, visiting Lima, Lake Titicaca and the mountain regions of the Andes.   Voyeikov sailed down the Amazon to Santarem. He intended to sail up the river Rio Negro and down the river Orinoco, but as the rainy season began, and with it an outbreak of yellow fever, he was forced to return to the mouth of the Amazon.   From there he reached New York by sailing vessel in February 1875.   During his travels around the American continent, Voyeikov master- ed another two languages, Spanish and Italian.

In June 1875 he returned home but in October of the same year he set off again, this time to Asia.   In India he visited Bombay, Calcutta, Benares and Delhi. Then, after spending six weeks on the island of Java, he crossed Southern China to Japan.   At that time European missionaries and merchants were only permitted to visit seven Japanese open ports and their environs, but Voyeikov as a scientist was allowed to travel everywhere in the country.   Accompanied by a young Japanese who knew Russian, Voyeikov visited the islands of Hokkaido, Honshu and Kyushu, covering a distance of more than 3,600 kilometres.   In Japan, as in America, his interest was primarily in the country's physico- geographical and climatic conditions.   Through ques- tions to the inhabitants and from the vegetation he was able to draw up the general outlines of a map of the climate of Japan, of which little had been known out- side that country until that time.   Voyeikov returned to St. Petersburg in January 1877.   He systematically published in various periodicals the observations he had made during his travels.

Voyeikov's world tour gave him a fine conception of its varied climates and helped him to define Russian climates in a discerning manner.   He travelled widely in Russia itself and in 1870 became the secretary of the Meteorological Committee set up by the Russian Geographical Society.   With his characteristic bubb- ling energy he organized a meteorological survey, especially of precipitation.   He tried to attract a large body of voluntary meteorological observers as he considered that in a country as vast as Russia scienti- fic generalization could only be sound if based on a mass of observations from individual stations, every one of which should be studied in its local geographi- cal setting.   From 1885 the government gave a grant

to support twelve modern meteorological stations, all of which Voyeikov visited annually.

At these stations the observers maintained an interest in agriculture for Voyeikov had an inherent understanding of the needs of farmers from his early years.   He was convinced that meteorological and climatic data could be of great value in agriculture. His dedication never lessened.   From 1883 he was chairman of the Meteorological Committee and from 1891 editor-in-chief of the new *Meteorological News*.   He carried both of these responsibilities to his death in 1916.

In 1884 he was elected Reader, in 1885 Professor extraordinary and in 1887 permanent Professor in St. Petersburg University.   In 1886 he travelled in Austria, Germany, Switzerland and Italy to study methods of teaching geography.   Finally, in 1915, he was elected as director of the first higher geograph- ical courses made available in Russia.   He attended the international congresses of the I.G.U. from 1881 onwards and was an honorary member or member of many Russian or foreign scientific societies.   Despite his fame, official recognition came late and only in 1910 was he elected a corresponding member of Russia's high- est scientific institution, the Academy of Sciences. But an abiding tribute was to name the main Geophysical Observatory in Leningrad after him.

2. SCIENTIFIC IDEAS AND GEOGRAPHICAL THOUGHT
Thanks to his exceptional capacity for work, his tre- mendous erudition, his wide scientific interests, his active response to the economic needs of society, and to his fluency in foreign languages, Voyeikov left a large scientific legacy of more than 1,700 works, including books, articles, abstracts of the works of other authors, reviews and notes.   Voyeikov's main work was the long monograph of 1884 *Klimaty zemnogo shara, v osobenosti Rossyi (Climates of the globe, with particular reference to Russia)*.   In this work he generalized the achievements of that time in the fields of meteorology, climatology, hydrology, and his own tremendous scientific experience.

Hann's famous *Handbuch der Klimatologie* was published in 1863 and remained a classic.   Voyeikov's own work of 1884 was published in German in 1887.   In it there was originality, for along with the actual descriptions of climates there was a full treatment of the reasons for climatic differences, an explanation of the essential character of meteorological phenomena and climatic processes and of the development of these processes in relation to other natural factors.   Long before his book appeared, Voyeikov had written exten- sively on the circulation of the earth's atmosphere. He showed that there were patterns of circulation not previously known and that climates were closely con- nected with atmospheric circulation.   Solar radiation was the motive force causing the air to circulate and a major responsibility of physical science was 'the introduction of an account book of the sun's warmth received by the globe with its air and water cover'.

Voyeikov was the first researcher to understand the influence of monsoons in areas beyond the Tropics. He discovered the ridge of high pressure formed over Asia in winter from Siberia to western Europe (known as 'Voyeikov's great mainland axis').   In his *Klimaty zemnogo shara (Climates of the globe)* he dealt with all the major climatic influences and showed that with solar radiation and atmospheric circulation moisture was of major significance. This led him to a detailed study of air humidity, evaporation, clouds, water precipitation, rivers and lakes. Precipitation and evaporation were inevitably opposing forces which determined the density of the river network and the condition of the rivers and lakes. To summarize:

> All conditions being equal, a country will be the richer in running water the more abundant the precipitation and the less the evaporation from the surface of the soil and water and from plants. Therefore rivers may be regarded as the product of climate. *(Selected works*, 1, 1948, 243.)

River flow was a major concern of Voyeikov. He showed that large rivers in the lower reaches of extensive drainage basins may exist independently of local climatic conditions. He examined several hydraulic and hydrographic aspects of river flow, including the effect on the runoff of soil permeability, the displacement of water along river beds and the stabilizing effect on river flow of lakes in their courses. This work led him to a classification of rivers according to their water supply. He recognized nine separate types of rivers according to climates. Normally rivers mirror the climate of the time but lakes reflect longer term changes in climate.

3. INFLUENCE AND SPREAD OF IDEAS

Voyeikov's contribution to geography was mainly as a climatologist at a time when knowledge of the world's climates was slowly advancing with the patient accumulation of local data, to which his work in Russia -- ultimately with government support -- was a significant contribution. The strength of his work, and in some ways its most crucial geographical quality, lay in his appreciation of the relation between climate and the physical environment, including soils and vegetation.

One major enterprise of Voyeikov was to calculate annual runoff of the earth's rivers, though in fact he underestimated it considerably. He made a pioneer calculation of the water balance in the Caspian sea which proved to be scientifically sound. He was also concerned with the problem of regular -- and probably also irregular -- changes in climate and opposed the view that central Asia was drying up progressively.

In his *Zomaty zemnogo shara (Climates of the globe)* he had much to say on snow. The cover of snow was a climatic factor of great significance, in part because snow reflects solar radiation into space. The snow cover mainly thaws out through the entry of air masses from warmer areas but in fact the heat balance was not a simple equation. Over the earth the heat resulting from the effect of snow far exceeds the cooling caused by it, for if there were no snow the world's dry cold lands would be more extensive than they in fact are. He also made various studies, with some interesting and original judgements on the climates of contemporary glacial regions and on those of the Ice Age.

Work on human, in his case chiefly economic, geography developed mainly from studying the climatic conditions of various regions of Russia. He put forward various suggestions on the agricultural crops that might be grown, many of which were followed, so that tea and citrus fruits were grown in Transcaucasia, cotton in central Asia, and flax and grain further north than before. And he also wrote on the climatic qualities of various Russian and foreign spas. Voyeikov might in modern times be regarded as an applied geographer or at least one whose work had obvious practical relevance in the agricultural life of Russia, with its need in some areas for drainage and its possibilities of irrigation in others, its particular problems of winter snows, spring thaw and varied soils. He was fortunate in possessing the financial resources to tour the world and so to make his fine contribution to climatology but his work on Russia was significant not only in his lifetime but in its redevelopment after his death in 1916, before the Revolution so long foreseen by the country's intelligentsia and others.

Bibliography and Sources

1. REFERENCES ON A. I. VOYEIKOV
Rikhter, G.D. 'Zhizn i Deyatelnost A.I. Voyeikova (The life and activity of A.I. Voyeikov)' in Voyeikov, A.I. *Izbranniye Sochineniya (Selected works),* Moscow (1948), vol 1, 35-82
Grigoriev, A.A. 'Rukovodyashchiye klimatologicheskiye idei A.I. Voyeikova (The leading climatological ideas of A.I. Voyeikov)', in Voyeikov, A.I. *Izbranniye Sochineniya (Selected works),* Moscow (1948), vol 1, 10-34
Markov, K.K. 'A.I. Voyeikov kak istorik climatov Zemli (A.I. Voyeikov as a historian of the climates of the Earth)', *Izv. Akad. nauk SSSR ser. geogr. (Proc. USSR Acad. geogr. ser.)',* (1951), no 3, 46-54
Pokshishevsky, V.V. (ed N.N. Baransky), 'A.I. Voyeikov kak ekonomiko -- geograf (A.I. Voyeikov as an economic geographer)' in *Otechestvenniye ekonomiki-geografy (Russian economic geographers),* Moscow (1957), 275-83

2. Principal works of A.I. Voyeikov
1870 'Uber des klima von Ost-Asien (on the climate of Eastern Asia)', *Z. Osterreich. Gesell. Meteor. (J. Austrian Meteor. Soc.),* vol 5, no 1, 39-42
1874 'Die atmosphärische Circulation (The circulation of the atmosphere)', *Pettermanns Mitt.,* Erganzungsband 8, 1873-4, no 38, 30p.: also in Russian in Voyeikov, A.I., *Izbranniye Sochineniya (Selected works),* vol 2, Moscow, 1949, 159-221
1875 'Raspredeleniye osadkov v Rossii (The distribution of precipitation in Russia)', *Zap. Russkogo Geogr. Obshch. po Obshchei Geogr. (Trans. Russian Geogr. Soc. Gen. Geogr.),* vol 6, book 1, 1-72
1877 'Puteshestviye po Yaponii, iyul -- oktyabr 1876g (Travels in Japan, July -- October 1876)', *Izv. Russkogo Geogr. Obshch. (Proc. Russian Geogr. Soc.),* vol 13, book 4, sect 2, 195-240

1879 'Kilmat oblasti mussonov Vostochnoi Asii (The
     climate of the Monsoon regions of East Asia)',
     Izv. Russkogo Geogr. Obshch., vol 15, book 5,
     sect 2, 321-410

1881 'Klimaticheskiye usloviya lednikovykh yavlenii
     nastoyashchikh i proshedshikh (The climatic con-
     ditions connected with present and past glacial
     phenomena)', *Zap. Mineral. Obshch. Trans.
     Mineral. Soc.)*, ser 2, part 16, 21-90

1884 *Klimaty zemnogo shara, v osobennosti Rossii
     (Climates of the Globe, with particular refer-
     ence to Russia)* in Voyeikov, A.I., *Izbranniye
     sochineniya (Selected works)*, Moscow (1948),
     161-750

1885 *Snezhniy pokrov, ego vliyaniye na klimat i pogodu
     i sposoby issledovaniya (The snow cover, its in-
     fluence on the climate and weather and ways of
     investigating it)*, St. Petersburg

1895 'O Klimate Tsentralnoi Asii na osnovanii
     nablyudenii chetyrekh ekspeditsii N.M. Przheval-
     skogo (On the climate of Central Asia on the basis
     of the observations made by N.M. Przhevalsky's
     four expeditions)' in Voyeikov, A.I. (ed),
     *Nauchniye resultaty puteshestvii Przhevalskogo po
     Tsentralnoi Asii (The scientific results of the
     travels of Przhevalski in Central Asia)*, St.
     Petersburg, 239-81

1899 'Klimat Polesya (The climate of the Polesiye
     area)' in Prilozheniya, K., *Ocherku rabot Zapadnoi
     ekspeditsii po osusheniya bolot za 1873-1898 gg
     (Appendix to an outline of the work of the
     Expedition on Swamp Drainage 1873-1898)*,
     St. Petersburg, 1-132

1903 *Meteorologia (Meteorology)*, parts 1-3,
     St. Petersburg, 556p.

1904 *Meteorologia (Meteorology)*, part 4,
     St. Petersburg, 210p.

1908 'Klimat Indiiskogo okeana i Indii (The climate of
     India and the Indian Ocean)', *Zap. po gidro-
     grafii (Bull. Hydrogr.)*, no 29, 178-263

1908 'Oroshenie Zakaspiiskoi oblasti s tochki zreniya
     geografii i climatologii (Irrigation of the
     Transcaspian area from the point of view of
     geography and climatology)', *Izv. Russ. Geogr.
     Obshch.*, vol 44, no 3, 131-60

1914 *Le Turkestan Russe*, Paris, 360p.

1915 'Klimaty russkikh i zagranichnykh lechebnykh
     mestnostei (Climates of Russian and foreign
     spas)', in *Prakticheskaya Meditsina (Practical
     medicine)*, no 6, 87-176 and no 10, 177-80 and
     Izbanniye sochineniya (Selected works), vol 1-4,
     Moscow, 1948-57 (ed. A.A. Grigoriev)

1949 *Vozdeistiviye cheloveka na prirodu, Izbranniye
     statii (Influence of man on nature, selected
     articles)*, Moscow, 1949, ed V.V. Pokshishoesky

*Dr. I.A. Fedosseyev is head of the section for the
history of geological and geographical sciences in the
USSR Academy of Sciences Institute of the History of
Natural Sciences and Technology, Moscow.*

Dates	Life and career	Activities, travel, fieldwork	Publications	Contemporary events and publications
1842	Born at Moscow			
1845				Organization of the Russian Geographical Society
1856–1858				Investigation of Asia by P.P. Semenov (Semenov-Tian-Shansky)
1860–1861	Studied at University of St. Petersburg			
1861–1865	Studied at Universities of Heidelberg, Berlin, Göttingen			Abolition of serfdom in Russia
1865	Presented his thesis at Göttingen University and became a Doctor of Philosophy		*Ueber die directe Insolation und Strahlung an vershiedenen Orten der Erdoberfläche* (Direct insolation in various parts of the Globe) (thesis)	
1866	Elected a member of the Russian Geographical Society			
1869		Visited the network of meteorological observatories in Western Europe		
1870	Became secretary of the Meteorological Commission of the Russian Geogr. Soc.		*Ueber das Klima von ost-Asien* (On the climate of eastern Asia)	Beginning of expeditions of N.M. Przhevalsky to Eastern and Central Asia (until 1885)
1872		Studied the chernozem soils in Galicia, Bukovina, Walachia, Transilvania and Austria		
1872–1875		Travelled through western Europe, United States and South America	*Die atmosphärische circulation* (The circulation of the atmosphere) (1874) *Raspredelenie osadkov v Rossii* (The distribution of precipitation in Russia) (1875)	
1975–1876		Travelled through India, Java, Southern China and Japan		
1877			*Puteshestviye po Japonii* (Travels in Japan) (1877)	

Dates	Life and career	Activities, travel, fieldwork	Publications	Contemporary events and publications
1881		Participated in the Int. Geogr. Congr. in Venice. Later he represented the Russian Geogr. Soc. at all other international geographical congresses.		
1883	Became the chairman of the Meteorological Commission of the Russian Geographical Society			
1884	Elected Reader in St. Petersburg University		*Klimaty zemnogo shara v osobennosti Rossii* (Climates of the globe, with particular reference to Russia)	Organization of chairs of geography at Russian universities
1885	Became Professor extraordinary		*Snezhniy pokrov, ego vliyaniye na klimat i pogodu i sposoby issledovaniya* (The snow cover, its influence on the climate and weather and ways of investigating it)	
1886– 1888		Travelled in Austria, Germany, Switzerland and Italy to study methods of teaching geography		
1887	Became permanent Professor			
1891	Became editor-in-chief of magazine *Meteorologicheskii vestnik* (Meteorological News)			
1895			*O Klimate Tsentralnoi Asii* (On the climate of Central Asia)	
1899				V.V. Dokutchaev, *Kucheniyu o zonah priordy* (On zones in nature)
1908			*Klimat Indiiskogo okeana i Indii* (The climate of India and the Indian Ocean)	
1910	Was elected corresponding member of the St. Petersburg Academy of Sciences			

Dates	Life and career	Activities, travel, fieldwork	Publications	Contemporary events and publications
1914			*Le Turkestan russe (Russian Turkestan)*	First World War
1915	Became director of the Higher Geographical Courses in St. Petersburg			Organization of Higher Geographical Courses in St. Petersburg
1916	Died in Petrograd (Leningrad)			

Index

The index is divided into three parts.

1. *PERSONAL NAMES* as far as possible are given in full for both the subjects and the authors of the individual studies, together with dates of birth (and death).
2. *ORGANIZATIONS AND RELATED REFERENCES* is subdivided under *Colleges, Institutes, Institutions, Museums, Official and Research Organizations; Scientific Congresses and Commissions; Societies* (including awards); and *Universities.*
3. *SUBJECTS* covers concepts, geographical theories and specific research. When a geographer included in the main sequence has made a specific contribution to a subject, his name is listed under the relevant entry in italics.

Page numbers in italic refer to the Bibliography and Sources sections of the individual biobibliographies and, similarly, underlined numbers refer to the chronological tables. In general the index to this volume follows that in volume 1, but with some modifications to section 2. It is hoped that these two indexes will be combined with those to future volumes of *GEOGRAPHERS* eventually to form a basis for research and inquiries in the history of geography and of geographical thought.

1. PERSONAL NAMES

2. ORGANIZATIONS AND RELATED REFERENCES

Colleges, Institutes, Institutions, Museums, Official and Research Organizations